Synthesis Lectures on Human Language Technologies

Series Editor

Graeme Hirst, Sandford Fleming Building, Department of Computer Science, University of Toronto, Toronto, ON, Canada

The series publishes topics relating to natural language processing, computational linguistics, information retrieval, and spoken language understanding. Emphasis is on important new techniques, on new applications, and on topics that combine two or more HLT subfields.

Serge Sharoff · Reinhard Rapp ·
Pierre Zweigenbaum

Building and Using Comparable Corpora for Multilingual Natural Language Processing

Springer

Serge Sharoff
Centre for Translation Studies
University of Leeds
Leeds, UK

Pierre Zweigenbaum
CNRS, Laboratoire Interdisciplinaire des
Sciences du Numérique
Université Paris-Saclay
Orsay, France

Reinhard Rapp
Faculty of Translation Studies, Linguistics
and Cultural Studies
University of Mainz
Germersheim, Germany

ISSN 1947-4040 ISSN 1947-4059 (electronic)
Synthesis Lectures on Human Language Technologies
ISBN 978-3-031-31383-7 ISBN 978-3-031-31384-4 (eBook)
https://doi.org/10.1007/978-3-031-31384-4

This Springer imprint is published by the registered company Springer Nature Switzerland AG
The registered company address is: Gewerbestrasse 11, 6330 Cham, Switzerland

Contents

About the Authors

Serge Sharoff is Professor of Language Technology and Digital Humanities at the Centre for Translation Studies, University of Leeds. Over the course of his academic career, he has published more than 150 peer-reviewed papers and supervised 18 Ph.D. students. His research focuses on Natural Language Processing, including automated methods for collecting very large corpora from the Web, their analysis in terms of domains and genres as well as extraction of lexicons and terminology from corpora. The application domains for this kind of research in the Digital Humanities include text annotation, information retrieval, machine translation and computer-assisted language learning. His research stresses the inherent multilingualism of NLP, which implies that tools and resources can be ported across languages by paying attention to the respective linguistic properties.

Reinhard Rapp is Professor of Applied Translation Studies at Magdeburg-Stendal University of Applied Sciences and is also affiliated with the University of Mainz. His main research interests are in computational linguistics, translation studies and cognitive science. Publications have dealt with unsupervised language learning from text corpora, word sense disambiguation, text mining, thesaurus construction, bilingual dictionary induction from parallel and comparable corpora, and with statistical and neural machine translation. He conducted EU-funded research projects at the University of Geneva, the University of Tarragona, the University of Leeds, at Aix-Marseille University, at the University of Mainz and at the Athena Research Center in Athens. He co-organized about 30 scientific conferences, workshops and shared tasks and is a co-editor of the book series 'Translation and Multilingual Natural Language Processing' (Language Science Press) and 'Linguistik International' (Peter Lang).

Pierre Zweigenbaum Ph.D., FACMI, FIAHSI, is a Senior Researcher at the Interdisciplinary Laboratory for Digital Sciences (LISN, Orsay, France), a laboratory of the French National Center for Scientific Research (CNRS) and Université Paris-Saclay, where he led the ILES Natural Language Processing group for seven years. Before CNRS he was a researcher at Paris Public Hospitals in an Inserm team for twenty years. He also was a part-time professor at the National Institute for Oriental Languages and Civilizations during ten years. His research focus is Natural Language Processing, with medicine as a main application domain. He has also designed methods to acquire linguistic knowledge automatically from corpora and thesauri, to help extend monolingual and bilingual lexicons and terminologies, using parallel and comparable corpora.

Introduction

1

1.1 Rationale for Working with Comparable Corpora

The ability of the computers to handle larger amounts of texts and the availability of more texts in electronic form led to the rise of data driven research in computational linguistics. In the case of Machine Translation (MT) and other kinds of multilingual Natural Language Processing (NLP), the first source of large data came from collections of translations, initially in the Statistical MT (SMT) approach from IBM [1], which was based on the Proceedings of the Canadian Parliament in English and French as their data source. This research direction was followed by a proliferation of SMT models, which relied on larger and larger collections of *parallel* data, which consist of exact translations between a pair of languages or several languages at the same time.

However, this early research also demonstrated the limits of using fully parallel corpora, with the most pressing initial concern on the amount of such data [4, 5]. This led to another strand of studies concerning the use of *less* parallel sources of texts, usually under the name of comparable corpora [6]. This book will present an overview of the modern approaches to building and using comparable data in multilingual NLP.

We will start by outlining the basic principles of using comparable resources as well as by comparing them to fully parallel resources (this chapter). This will be followed by specific chapters on building comparable corpora (Chap. 3), aligning their sentences to find a database of suitable translations (Chap. 4), using these corpora to produce dictionaries and termbanks (Chap. 5), to build MT engines (Chap. 6) and to use them in other applications (Chap. 7).

S. Sharoff et al., *Building and Using Comparable Corpora for Multilingual Natural Language Processing*, Synthesis Lectures on Human Language Technologies,
https://doi.org/10.1007/978-3-031-31384-4_1

1.1.1 Availability of Truly Parallel Data

The first limitation in the use of parallel corpora comes from the process of their production. For truly parallel corpora we need to collect texts that have been carefully translated by highly trained professional translators [8]. Many more people produce monolingual texts in their native languages in comparison to output of a small number of trained translators. Also there is an imbalance in the amount of translations produced for a relatively small number of major languages, primarily for the languages of the United Nations or the EU, while there are thousands of languages with very minor resources. Less resourced languages can still benefit from parallel resources. However, statistically speaking, language can be described as a large number of rare events in the sense that an individual word or expression is relatively infrequent, but the totality of rare events contributes to the mass probability of words and expressions in a text [9]. This creates problems of sparsity even for better-resourced languages. For example, the word *unicyclist* occurs 84 times in 2 billion words of a monolingual English ukWac corpus [10], i.e., about once per 23 million words in ukWac, with no examples of this word in the English parts of large publicly available parallel corpora such as Europarl [15] or the United Nations corpus [16].

Another constraint on corpus data concerns the availability of translation products, as it is easier to obtain translated texts produced by large public bodies, such as the European Parliament or the United Nations than many other kinds of translations. This bias leads to many word choices, which are specific with respect to the genres and topics available in such corpora. For example, there are 75 occurrences of the expression *strong voice* in Europarl, all of which are used in the sense of political authority, for example, *ensuring that smaller Member States retain a* ***strong voice*** *in the decision-making procedures*. At the same time, out of the 28 occurrences of this expression in the British National Corpus [17] 19 examples refer to the quality of human voices, for example, *She had a good,* ***strong voice****—an actor's voice*. When translating *strong voice* into other languages, the political authority metaphor needs to be explicitly unpacked, because the literal translation is not suitable. For example, when translating *strong voice* into Russian, the political sense is likely to use *reshitelno vystupatj* ('to express assertively') or *goryacho osuzhdat* ('to condemn vehemently') as opposed to the straightforward literal translation *gromkij golos* ('loud voice'). In the end, the mismatch between the domains and genres of parallel corpora and the target applications can lead to errors, which might be corrected by the use of less parallel resources.

1.1.2 Translationese in Parallel Data

Another kind of problems concerning the use of parallel data comes from a particular phenomenon known as translationese, namely a difference between features of translated texts and texts originally produced by native speakers [18]. Translationese is caused by factors inherent in the translation process, such as explicitation [19], i.e., the need to provide more information in the translated text for what remains implicit in the source text. For example, translations tend to use more cohesive markers, such as *therefore, however, nevertheless*, in comparison to original texts by making the logical relations more explicit in translation [18]. Other factors leading to translationese are related to the temporal and cognitive pressures on the translation operations, as the translators need to complete their tasks in a short period of time. This leads such phenomena as (1) normalisation, i.e., the tendendency to re-use stock expressions of the target language even when the source text deviates from the norm, and (2) "shining-through" [20], i.e., the influence of the syntactic and lexical choices stemming from the source texts even when other choices are common in monolingually produced texts in the target languages. In the end it has been shown that translationese effects have statistical significance leading to the ability to build fairly accurate classifiers detecting texts translated by humans [21], even in an unsupervised fashion [22].

From the viewpoint of the translation direction, there is always a source text which needs to be translated into other languages, so only one text in a parallel corpus is primary, while other texts exhibits features of translationese. At the same time, when parallel corpora are used for MT applications, this directionality of translation is usually ignored, so that the Slovenian→ English MT built from the Europarl corpus uses slightly unnatural Slovenian texts exhitibing features of translationese as its source texts. These aspects call for the greater use of texts originally produced in the respective languages.

1.2 Levels of Comparability

These constraints on the availability and the use of parallel data led to numerous studies utilising less parallel resources. This research direction generally goes under the name of 'comparable corpora'. However, it is important to note that there is no clear dividing line between fully parallel and comparable corpora, as multilingual resources vary with respect to the degree of linking between documents in the two languages. Consider some examples of documents along the cline of comparability:

translations identifiable originally produced source texts and their translations.

- **truly parallel** true high quality translations, such as the proceedings of the European Parliament.
- **modified parallel** translations with some modifications to cater for the target audience. For example, language-specific descriptions of the Search dialogue box in the OpenOffice manual are translations from English with necessary modifications, such as searching for *New York* is replaced with searching for *Berlin* in the German version.
- **adaptations** translators exhibit freedom in rendering the source text, as it is the case with many of the fan-produced subtitles as in the OpenSubtitles corpus [23].

Table 1.1 Adena culture example

en	The Adena culture was a Pre-Columbian Native American culture that existed from 800 BC to 1 AD, in a time known as the Early Woodland period
de	Adena-Kultur ist die Bezeichnung für eine im mittleren Ohio-Tal ansässige prähistorische Indianerkultur. Sie lässt sich für die Zeit von etwa 1000 v. Chr. bis 200 n. Chr. nachweisen
fr	La Civilisation Adena était une culture pré-colombienne amérindienne ayant existé de l'an 1 000 à l'an 200 avant J.-C., durant l'ère connue sous le nom de Période sylvicole
ja	アデナ文化（Adena）は、アメリカ合国オハイオ州を中心に1000B.C.から元前後にえた文化。アデナ文化は、初期（ないし前期）ウッドランド期（Early Woodlannd Period）の文化として位置付けられ、アデナ文化の出によって、後のホプウェル文化をはじめとするウッドランド文化の先となるウッドランド式土器や丘墓、トウモロコシ耕のすべてがでそろった。
ru	Культура Адена - доколумбовая индейская археологическая культура, существовавшая в период 1000-200 г. до н. э., в период, известный как ранний Будлендский период.

strongly comparable closely related texts produced in several languages.

Wikipedia entries entries on exactly the same topic are linked across languages via iWiki links, see Table 1.1, with individual entries varying in the amount of information.

news items very specific events are covered by numerous news agencies in various languages. Often the same agency reports the same story in various languages without fully relying on their translation, see the BBC News in English and Spanish.

weakly comparable similar texts which cannot be directly linked to each other across languages, while still being in the same domain and genre, for example:

- texts in the same narrow subject domain and genre, but describing different events, e.g. parliament debates on health care from the German Bundestag, the British House of Commons and the French parliament;
- texts within the same broader domain and genre, but varying in subdomains and specific genres, e.g. parallel queries in the renewable energy domain mostly returning wind energy research articles in English vs solar panel producers in Russian [24].

unrelated collections of unrelated texts, which are nevertheless collected using comparable methods from comparable sources. For instance, this concerns the use of random snapshots of the Web for Chinese, English, German and Russian [25] or the use of filtered Common Crawl data [26] with the assumption that different cultures use the Web for broadly similar purposes.

Further on in the book we will refer to collections of documents on this comparability scale as parallel, strongly comparable, weakly comparable and unrelated corpora.

1.3 Methodology for Dealing with Comparable Resources

Even though there is no clear dividing line between parallel and comparable resources, there are two clearly distinct ways of using such resources, which focus on the amount of translation information present in each resource.

The first approach assumes the possibility of obtaining very local information which can link two languages through splitting corpora into small translation units, which are supposed to convey the same meaning in both languages. Usually this is done on the sentence level. This is the classic case of training Machine Translation (MT) engines, as MT naturally assumes a high level of parallelism as the starting point. This can be also considered as a fully supervised model of translation.

The second approach assumes building global multi-dimensional spaces for the entire set of linguistic phenomena in each language to determine the similarity between them. This is the classic case of using comparable resources for dictionary induction, the phenomena in this case are words or other kinds of dictionary entries. As usual in computational linguistics there are many studies blending the local and global approaches.

Irrespectively of choosing the local or global approach to using corpora, we need to determine ways of measuring similarity. Modern approaches usually rely on the Vector Space Model (VSM), which uses a numerical vector x to represent any entity, such as a word, a phrase, a sentence or a link between entities. If a VSM assign similar vectors to entities, which are translations of each other in the respective languages, this shared space can be the basis for identifying translations via comparable corpora. We will start with an introduction into the basic principles of building cross-lingual VSMs, see Chap. 2, and then we will proceed to methods for utilising these spaces, first for finding similar documents, see Chap. 3, then for finding translations on the sentence level, see Chap. 4 and finally for bilingual dictionary induction on the word level, Chapter 5. The shared VSM space is also a natural starting point for developing Machine Translation engines with no or with limited parallel data, see Chap. 6.

The utility of comparable corpora goes beyond applications in translation. As the amount of resources required for training NLP tools for text annotation, document classification or retrieval is limited in many languages, it is possible to achieve low-resource NLP via few-shot or zero-shot models. This means an NLP model can be trained for a low-resource language with very few or even no annotated examples in this language provided that we have a sufficient number of annotated examples in another (donor) language and that we have comparable corpora for both languages. Methods for making cross-lingual predictions in this context are discussed in Chap. 7.

References

1. Brown PF, Cocke J, Della Pietra SA, Della Pietra VJ, Jelinek F, Lafferty JD, Mercer RL, Roossin PS (1990) A statistical approach to machine translation. Comput Linguist 16(2):79–85
2. Buck C, Koehn P (2016) Findings of the WMT 2016 bilingual document alignment shared task. In: Proceedings of the first conference on machine translation: volume 2, shared task papers. Association for Computational Linguistics, Berlin, Germany, pp 554–563. https://doi.org/10.18653/v1/W16-2347, https://aclanthology.org/W16-2347
3. Cai X, Huang J, Bian Y, Church K (2021) Isotropy in the contextual embedding space: clusters and manifolds. In: International conference on learning representations
4. Rapp R (1995) Identifying word translations in non-parallel texts. In: Proceedings of ACL, Cambridge, pp 320–322
5. Fung P (1998) A statistical view on bilingual lexicon extraction: from parallel corpora to non-parallel corpora. In: Machine translation and the information soup. Springer, pp 1–17. http://www.springerlink.com/content/pqkpwpw32f5r74ev/
6. Sharoff S, Rapp R, Zweigenbaum P (2013) Overviewing important aspects of the last twenty years of research in comparable corpora. In: Sharoff S, Rapp R, Zweigenbaum P, Fung P (eds) BUCC: building and using comparable corpora. Springer, pp 1–17
7. Sharoff S, Zweigenbaum P, Rapp R (2015) BUCC shared task: cross-language document similarity. In: Proceedings of eighth workshop on building and using comparable corpora. Association for Computational Linguistics, Beijing, China, pp 74–78
8. Massey G (2017) Translation competence development and process-oriented pedagogy. In: The handbook of translation and cognition. Wiley Online Library, pp 496–518
9. Harald B (2008) Analyzing linguistic data. Cambridge University Press, Cambridge
10. Baroni M, Bernardini S, Ferraresi A, Zanchetta E (2009) The WaCky wide web: a collection of very large linguistically processed web-crawled corpora. Lang Resour Eval 43(3):209–226
11. Baroni M, Dinu G, Kruszewski G (2014) Don't count, predict! A systematic comparison of context-counting versus context-predicting semantic vectors. In: Proceedings of the 52nd annual meeting of the association for computational linguistics, Baltimore, Maryland
12. Bateman JA (1997) Enabling technology for multilingual natural language generation: the KPML development environment. Nat Lang Eng 3(1):15–55
13. Bengio Y (2012) Deep learning of representations for unsupervised and transfer learning. In: Guyon I, Dror G, Lemaire V, Taylor G, Silver D (eds) Proceedings of ICML workshop on unsupervised and transfer learning. Volume 27 of proceedings of machine learning research, Bellevue, Washington, USA. PMLR, pp 17–36. https://proceedings.mlr.press/v27/bengio12a.html. Accessed 02 July 2012
14. Bernardini S, Ferraresi A (2013) Old needs, new solutions: comparable corpora for language professionals. In: Sharoff S, Rapp R, Zweigenbaum P, Fung P (eds) Building and using comparable corpora. Springer, pp 303–319
15. Philipp K (2005) Europarl: a parallel corpus for statistical machine translation. In: Proceedings of MT summit X
16. Ziemski M, Junczys-Dowmunt M, Pouliquen B (2016) The United Nations parallel corpus v1.0. In: Proceedings of the tenth international conference on language resources and evaluation (LREC'16), Portorož, Slovenia, May 2016. European Language Resources Association (ELRA)

17. Guy A, Lou B (1998) The BNC handbook: exploring the British national corpus with SARA. Edinburgh University Press, Edinburgh
18. Koppel M, Ordan N (2011) Translationese and its dialects. In: Proceedings of the 49th annual meeting of the association for computational linguistics: human language technologies, Portland, Oregon, USA, June 2011. Association for Computational Linguistics, pp 1318–1326. https://www.aclweb.org/anthology/P11-1132
19. Frankenberg-Garcia A (2009) Are translations longer than source texts. In: Beeby A, Rodriguez P, Sanchez-Gijon P (eds) Corpus use and learning to translate (CULT): an introduction. John Benjamins, pp 47–58
20. Rubino R, Lapshinova-Koltunski E, van Genabith J (2016) Information density and quality estimation features as translationese indicators for human translation classification. In: Proceedings of the 2016 conference of the North American Chapter of the association for computational linguistics: human language technologies, San Diego, CA, June 2016. Association for Computational Linguistics, pp 960–970. https://doi.org/10.18653/v1/N16-1110. https://www.aclweb.org/anthology/N16-1110
21. Ella R, Shuly W (2015) Unsupervised identification of translationese. Trans Assoc Comput Linguist 3:419–432
22. Riley P, Caswell I, Freitag M, Grangier D (2020) Translationese as a language in "multilingual" NMT. In: Proceedings of the 58th annual meeting of the Association for Computational Linguistics. Association for Computational Linguistics, pp 7737–7746. https://www.aclweb.org/anthology/2020.acl-main.691
23. Tiedemann J (2012) Parallel data, tools and interfaces in OPUS. In: Proceedings of the eighth international conference on language resources and evaluation (LREC'12), Istanbul, Turkey, May 2012. European Language Resources Association (ELRA), pp 2214–2218. http://www.lrec-conf.org/proceedings/lrec2012/pdf/463_Paper.pdf
24. Sharoff S (2013) Measuring the distance between comparable corpora between languages. In: Sharoff S, Rapp R, Zweigenbaum P, Fung P (eds), BUCC: building and using comparable corpora. Springer, Berlin/New York, pp 113–129
25. Sharoff S (2006) Open-source corpora: using the net to fish for linguistic data. Int J Corpus Linguist 11(4):435–462
26. Wenzek G, Lachaux M-A, Conneau A, Chaudhary V, Guzman F, Joulin A, Grave E (2019) CCNET: extracting high quality monolingual datasets from Web crawl data. arXiv:1911.00359
27. Wu D (1994) Aligning a parallel English-Chinese corpus statistically with lexical criteria. In: Proceedings of the 32nd annual meeting on Association for Computational Linguistics (ACL '94), Stroudsburg, PA, USA, 1994. Association for Computational Linguistics, pp 80–87
28. Wu S, Dredze M (2019) Beto, Bentz, Becas: the surprising cross-lingual effectiveness of BERT. arXiv:1904.09077
29. Wu S, Dredze M (2020) Are all languages created equal in multilingual BERT? In: Proceedings of the 5th workshop on representation learning for NLP, Online, July 2020. Association for Computational Linguistics, pp 120–130. https://doi.org/10.18653/v1/2020.repl4nlp-1.16. https://aclanthology.org/2020.repl4nlp-1.16
30. Xiao R (ed) (2020) Using corpora in contrastive and translation studies. Cambridge Scholars Publishing
31. Xu H, Koehn P (2021) Cross-lingual BERT contextual embedding space mapping with isotropic and isometric conditions. arXiv:2107.09186

Basic Principles of Cross-Lingual Models 2

When we start working across languages, we need to determine ways of measuring similarity of entities (such as a word, a phrase, a sentence or a link between entities) within each language and across languages. Modern approaches usually rely on the Vector Space Model (VSM), which uses a numerical vector x of a specified dimensionality D to represent any entity. The entities are similar when their vectors are not too far apart.

2.1 Monolingual VSMs

Early VSM approaches [1] constructed these vectors in an intuitive model by computing the strength of association between entities. For example, a word can be represented as a vector of other words it co-occurs with. The values of these vector can be either normalised frequencies of co-occurences or some kinds of collocation scores in a window of a certain size [2]. Table 2.1 lists example vectors for three words *moon, Sun* and *zebra.*[1]

Each row of this matrix **M** stores the Point-wise Mutual Information scores [4] for the collocates:

$$PMI = \log \frac{p(w_1, w_2)}{p(w_1)p(w_2)} \tag{2.1}$$

where p is the probability of seeing words w_1, w_2 or their collocation "$w_1 w_2$" in this corpus. The PMI value is assumed to be zero if $p(w_1, w_2)$ is zero, i.e. if the two words do not

[1] The frequencies have been taken from the British National Corpus [3], which was compiled in the beginning of the 1990s. This timeframe leads to the specific *Sun-IBM* collocation.

S. Sharoff et al., *Building and Using Comparable Corpora for Multilingual Natural Language Processing*, Synthesis Lectures on Human Language Technologies,
https://doi.org/10.1007/978-3-031-31384-4_2

Table 2.1 Examples of word vectors in an association matrix

	crescent	cross	full	giraffe	half	hot	IBM	moon	rise	see	shine	stripe	sun
Moon	9.3	0.0	8.2	0.0	7.9	0.0	0.0	0.0	6.8	3.3	7.6	0.0	10.5
Sun	0.0	0.0	5.3	0.6	0.8	8.7	8.3	10.5	8.1	4.1	11.1	0.0	0.0
Zebra	0.0	13.6	0.0	9.3	0.0	0.0	0.0	0.0	0.0	0.7	0.0	11.8	0.0

co-occur in a specific window, the distance of three words in either direction was chosen for computing this table.

This similarity of vectors implies the similarity of meanings according to the hypothesis of distributional similarity, *You shall know a word by the company it keeps*, as originally formulated by Firth [5], see also Harris [6], Turney and Pantel [7], i.e., semantically similar words are likely to have similar vectors of co-occurences.

A VSM representation of this kind is interpretable, as one can see how words are related to each other via their collocations, so that we can see the conditions under which words are similar, for example, we can see that *moon* and *sun* share a number of collocates, and under which they differ, because the collocations apply to one of the senses but not to others. For example, some collocations are specific to one sense of *sun* (*hot, moon, rise, shine*), while others are specific to another sense, such as its collocation with *IBM* for the computer company sense of *Sun*, as in the BNC example *offered neither by Sun nor IBM*.

Even though the model is easily interpretable, its application creates a problem with the number of dimensions in the co-occurrence vector, because the vector for each word needs to include *all* words selected as possible collocates for every other word in this corpus. This becomes intractable in a large corpus with a lexicon easily exceeding several million words, while the matrix of collocations will be the square of this size. Depending on how aggressively one reduces the number of the most important keywords (this defines D, the dimensionality of each row in matrix $\mathbf{M}$), the number of keywords in a large corpus is nevertheless likely to exceed 100,000. This is still not practical, as there are many issues associated with high-dimensional spaces, such as the curse of dimensionality and hubness, when some entities are more likely to become closer to a large number of other entities, thus polluting the usefulness of similarity estimates. The likelihood of such problems rises with the number of dimensions [8]. At the same time, the matrix is very sparse, as most of its values are zeros. For example, *zebra* does not co-occur with most of the collocates of *moon* and *sun*, while it adds its own set of collocates *giraffe*, *stripe*, and *cross* (the last collocate comes from the sense of *zebra crossing*). In the end, the vector for *zebra* is similar to vectors for other animals and their properties (*giraffe* and *stripe*), as well as to various concepts related to the transport infrastructure (from its sense of *zebra crossing/*), which in turn have very few collocations shared with *moon*.

A VSM space can be compressed to a denser representation by keeping only a smaller number of dimensions $d \ll D$ which still retains most of the variation available in the full matrix of co-occurrences. A simple way to achieve this is by using the Singular Value Decomposition of matrix $\mathbf{M} = \mathbf{U}\mathbf{\Sigma}\mathbf{V}^T$ [9] to derive a low-rank approximation matrix $\mathbf{M}_d$ such that:

$$\mathbf{M}_d = \mathbf{U}\mathbf{\Sigma}_d\mathbf{V}^T \tag{2.2}$$

in which only the largest d singular values of $\mathbf{\Sigma}$ are retained. The commonly used value of d is between 50 and 500. Another popular approach for building dense VSM spaces is Random Semantic Indexing (RSI), see Sahlgren et al. [10], which also avoids the problem of making a direct SVD transform on a co-occurrence matrix with millions of dimensions. Performing either SVD or RSI produces non-sparse vectors of a reasonable number of dimensions, which simplifies many operations in this vector space. However, the downside is that the resulting dimensions no longer interpretable.

The neural approach to building VSMs is similar to the collocation-based approach except that it replaces the need of counting the co-occurrences with predicting their probability. If we take a very large set of examples in which one word has been removed, such as

Example 2.1 *shine of the .... upon the harbour*

Example 2.2 *light of the .... shining between clouds*

it is possible to recover the most likely masked word with a certain degree of precision, for example, for (2.1) and (2.2) this is likely to be either *moon* or *sun*. Therefore, we can learn dense vector representations (called embeddings in the neural approach) which assign similar vectors to words which can be used in similar contexts. For example, the initially random vectors for *sun* and *moon* through this training are becoming similar, because their embedding vectors are trained to predict similar contexts, such as *full, rise* or *shine*. From a random initialisation state, a neural network is trained to learn more suitable embedding vectors for the respective words with backpropagation from the masked word prediction task. For the best results we need to adjust the schedule of presenting words (to balance the frequency effects) and to choose the rate of negative examples, i.e. words not occurring in the chosen context by randomly inserting them into the empty slot. Randomly chosen words will either violate grammatical constraints or offer unreasonable contexts:

Example 2.3 *shine of the APPLY\BLOOD upon the harbour*

More formally, we can build word embedding representations by modeling the probability of seeing a word in the context of m other words:

$$p(w, c_{1:m}) = \frac{1}{1 + e^{-\sum w \cdot c_i}}$$

where the function is the sigmoid, which assumes the probability close to 1 for real corpus examples and the probability close to 0 for negative examples. We can find the best vectors w and c to match the probabilities for the positive and negative examples by using the Stochastic Gradient Descent algorithm, see Chap. 10 in Goldberg [11] for more information on the training process. Learning representations rather than counting co-occurrences usually leads to better generalisation capabilities of the neural embedding models in comparison to the count-based ones [12].

The beauty of neural methods is that embedding representations can be learned on the levels of words, sentences, paragraphs or documents, thus providing vector representations, which are suitable for many downstream tasks, so that embeddings can be used for Machine Translation, text classification, natural language inference or part-of-speech tagging. The objective of learning suitable embeddings aims at producing similar vectors (i.e. close with respect to their distance in the VSM space) for words, sentences, etc., when they are used in similar contexts, therefore similar embedding vectors are assumed to be similar in meanings. The common practice is to use embedding VSMs with around 300 dimensions for word embedding models (same as with SVD or Random Semantic Indexing), rising to around 1000 dimensions for sentence or paragraph embedding models.

2.2 Cross-Lingual VSMs

Creating VSMs that can encode multiple languages is very useful in a multilingual context. The VSM spaces can be either built in a way which creates them shared across several languages, or two VSM spaces from unrelated comparable corpora can be aligned post-hoc, for example, by using a seed dictionary or merely identical tokens, such as numbers or punctuation marks. In the case of post-hoc alignment, the two independent VSM representation spaces are transformed with the objective of allocating the seed dictionary elements as close as possible in this aligned space.

We can use comparable corpora in building VSM representationss if we can find a way of aligning the VSMs for several languages, which means obtaining similar vector representations for similar entities (words, constructions or sentences) **across** languages. Figure 2.1 shows word vectors for several English words (*moon, sun, giraffe, zebra*) and the related French words (*lune, soleil, girafe, zèbre*) according to the aligned word embeddings of fasttext [13], as obtained after reducing the dimensionality of the $d = 300$ fasttext vectors to a two-dimensional plot via a t-SNE transformation [14]. We can determine that the closest French word to *sun* in this space is *soleil*, which helps in inferring their semantic similarity.

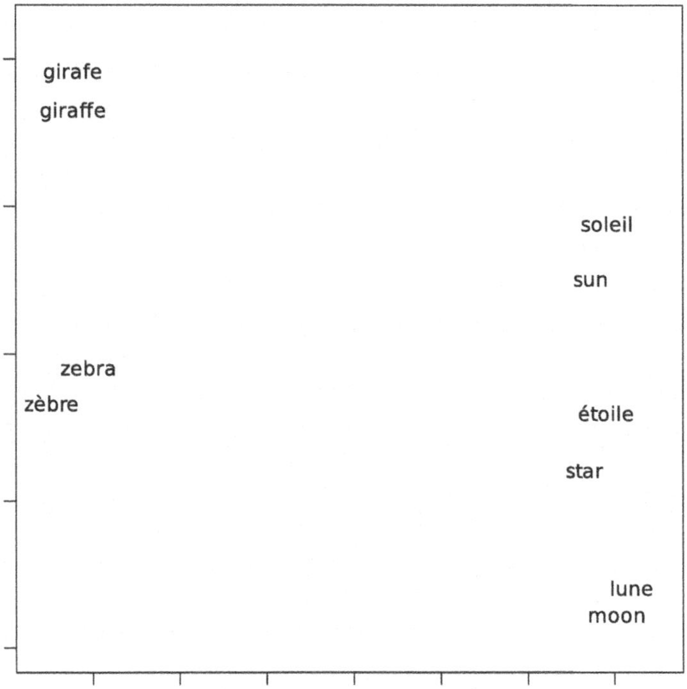

Fig. 2.1 Example of aligned word spaces

2.3 Contextual Embeddings

One of the problems with traditional vector space models is the need to represent the meaning of each word as a single vector. Such examples as *sun* and *zebra* show that the assumption that words with similar vectors have similar meanings is often difficult to satisfy. Many words have very different meanings in different contexts, so that the embedding for *sun* needs to be close to both *moon* and *IBM* at the same time, while this is not true for its likely French translation *soleil*.

A way of representing the way how the embeddings for words are combined together to produce meanings in their contexts is offered by means of connected layers of mathematical objects called artificial neurons (they are called neurons because their design was inspired by biological neurons, even though the objects themselves and their networks are usually quite different from the biological neurons). Three relevant architectures here are (1) Recurrent Neural Networks (RNNs), which model how much the embedding for the next word adds to the embedding vector obtained from the preceding words, (2) Convolutional Neural Networks (CNNs), which model patterns of co-activation of word embeddings irrespectively of their order, and (3) Transformers, which model the so called attention mechanism, i.e. how words influence each other. For example, for a sentence like.

Example 2.4 *The **food** was **amazing**, **but** the **service left** something to be **desired**.*

The transformer builds attention links between words, such as the objects and their properties, e.g.,*food* and its quality. The weights for combining word embeddings in all of these architectures are estimated by means of back-propagation of errors in a classification task, i.e., the weights are optimised to reduce the classification loss errors using such algorithms as Stochastic Gradient Descent. The principles behind neural architectures and their training are introduced in far greater details in Goldberg [11].

Contextual embedding models, such as ELMO or BERT, combine the mechanism of training word embeddings to predict masked words, such as in Examples (2.1) or (2.2) with the mechanism of neural networks which model how words can influence each other. For example, in the case of BERT, the transformer model is pre-trained on a large corpus (such as Wikipedia) to predict masked words or whether a sequence of two sentences is real or not. This pre-trains the weights for attention links from a very large raw text corpus without any explicit annotation. The attention links which can be detected as the outcome of this translating can already predict many kinds of syntactic and semantic relations, for example, subject-verb agreement, even without any explicit training [15]. The weights can be later "fine-tuned" to a downstream task by training on a downstream corpus, for example, the sentiment of the whole sentence or evaluation of its specific properties, such as food or service [16].

The weights in each layer of the pre-trained transformer models determine how the embeddings for individual words are modified dynamically for each example. In the end, the embedding for *Sun* in the upper layer of this architecture when used in a context like *offered neither by Sun nor IBM* will be different from the embedding the same token in the same layer when used in a context like *alignment of the Sun, Moon and Earth*. The pre-trained transformer models started with BERT [17] and this was later followed by a family of models, such as Roberta [18] or Distilbert [24].

Even in the absence of truly parallel corpora or any dictionary seeds (as used for producing the representations like Fig. 2.1, pre-trained language models can achieve a good degree of parallelism through unsupervised methods, as they detect inherent parallelism in the distribution of frequencies of words and constructions via sharing the weight parameters [25, 26]. For example, if we take Wikipedia corpora for English and French and set the task of predicting the masked words for a number of sentences like.

Example 2.5 *Consult the TABLE beam sizes below* versus
Vous pouvez consulter le TABLEAU des rémunérations des professeurs
'you can consult the TABLE of the salaries of the teachers'
where the capitalised *TABLE* and *TABLEAU* are the masked words, the resulting embedding representations for many words in the lexicon will be sufficiently close to their relevant translations in another language, even though the sentences themselves are not parallel translations.

In the remaining chapters of this book, we will show how representations of this kind are useful to predict the similarity of documents, sentences and words across languages, as well as to build MT and other applications.

We want to stress here that the majority of methods for building multilingual VSMs rely on comparable corpora with the most successful recent methods relying on unrelated comparable corpora. Nevertheless, to understand which models are more successful than others we need to understand how comparable their pre-training corpora are. For example, the multilingual BERT is trained on Wikipedias, while XLM-Roberta is trained on the multilingual Common Crawl corpus [27]. Sharing parameters in their pre-training comes from the comparability of their underlying corpora, which we are going to investigate in the forthcoming chapters.

References

1. Salton G, Wong A, Yang C-S (1975) A vector space model for automatic indexing. Commun ACM 18(11):613–620
2. Manning C, Schütze H (1999) Foundations of statistical natural language processing. MIT Press, Cambridge, MA
3. Aston G, Burnard L (1998) The BNC handbook: exploring the British national corpus with SARA. Edinburgh University Press, Edinburgh
4. Church K, Hanks P (1990) Word association norms, mutual information, and Lexicography. Comput Linguist 16(1):22–29
5. Firth JR (1951) Modes of meaning. In: Firth JR (ed) Papers in linguistics 1934–1951. Oxford University Press, Oxford, 1957, pp 190–215
6. Harris Z (1954) Distributional structure. Word 10(23):146–162
7. Turney PD, Pantel P (2010) From frequency to meaning: vector space models of semantics. J Artif Intell Res 37:141–188
8. Radovanović M, Nanopoulos A, Ivanović M (2010) Hubs in space: popular nearest neighbors in high-dimensional data. J Mach Learn Res 11(Sep):2487–2531
9. Rapp R (2004) A freely available automatically generated thesaurus of related words. In: Proceedings of the language resources and evaluation conference, (LREC'04), Lisbon, pp 395–398
10. Sahlgren M, Holst A, Kanerva P (2008) Permutations as a means to encode order in word space. In: Proceedings of the 30th annual meeting of the cognitive science society, Washington DC, July 2008
11. Goldberg Y (2017) Neural network methods for natural language processing. Synthesis lectures on human language technologies. Morgan & Claypool Publishers
12. Baroni M, Dinu G, Kruszewski G (2014) Don't count, predict! A systematic comparison of context-counting versus context-predicting semantic vectors. In: Proceedings of the 52nd annual meeting of the Association for Computational Linguistics, Baltimore, MD, June 2014
13. Bojanowski P, Grave E, Joulin A, Mikolov T (2017) Enriching word vectors with subword information. Trans Assoc Comput Linguist 5:135–146
14. Van der Maaten L, Hinton G (2008) Visualizing data using t-SNE. J Mach Learn Res 9:2579–2605
15. Rogers A, Kovaleva O, Rumshisky A (2020) A primer in BERTology: what we know about how BERT works. Trans Assoc Comput Linguist 8:842–866

16. Pontiki M, Galanis D, Papageorgiou H, Androutsopoulos I, Manandhar S, AL-Smadi M, Al-Ayyoub M, Zhao Y, Qin B, De Clercq O, Hoste V, Apidianaki M, Tannier X, Loukachevitch N, Kotelnikov E, Bel N, iménez-Zafra SM, Eryiğit G (2016) SemEval-2016 task 5: aspect based sentiment analysis. In: Proceedings of the 10th international workshop on semantic evaluation (SemEval-2016), San Diego, CA, June 2016. Association for Computational Linguistics, pp 19–30. https://doi.org/10.18653/v1/S16-1002. https://www.aclweb.org/anthology/S16-1002
17. Devlin J, Chang M-W, Lee K, Toutanova K (2019) BERT: pre-training of deep bidirectional transformers for language understanding. In: Proceedings of the 2019 conference of the North American chapter of the Association for Computational Linguistics: human language technologies, vol 1 (Long and Short Papers), Minneapolis, MN, June 2019. Association for Computational Linguistics, pp 4171–4186. https://aclanthology.org/N19-1423
18. Liu Y, Ott M, Goyal N, Du J, Joshi M, Chen D, Levy O, Lewis M, Zettlemoyer L, Stoyanov V (2019) RoBERTa: a robustly optimized BERT pretraining approach. arXiv:1907.11692
19. Lyu S, Son B, Yang K, Bae J (2020) Revisiting modularized multilingual NMT to meet industrial demands. In: Proceedings of the 2020 conference on empirical methods in natural language processing (EMNLP), Online, Nov 2020. Association for Computational Linguistics, pp 5905–5918. https://doi.org/10.18653/v1/2020.emnlp-main.476. https://aclanthology.org/2020.emnlp-main.476
20. Massey G (2017) Translation competence development and process-oriented pedagogy. In: The handbook of translation and cognition. Wiley Online Library, pp 496–518
21. McCann B, Bradbury J, Xiong C, Socher R (2017) Learned in translation: contextualized word vectors. arxiv:1708.00107
22. Melamed ID (1999) Bitext maps and alignments via pattern recognition. Comput Linguist 25(1):107–130
23. Melamed ID (2000) Models of translation equivalence among words. Comput Linguist 26(2):221–250. https://aclanthology.org/J00-2004
24. Sanh V, Debut L, Chaumond J, Wolf T (2019) DistilBERT, a distilled version of BERT: smaller, faster, cheaper and lighter. arXiv:1910.01108
25. Pires T, Schlinger E, Garrette D (2019) How multilingual is multilingual BERT? arXiv:1906.01502
26. Conneau A, Wu S, Li H, Zettlemoyer L, Stoyanov V (2020) Emerging cross-lingual structure in pretrained language models. In: Proceedings of the 58th annual meeting of the Association for Computational Linguistics, Online, July 2020. Association for Computational Linguistics, pp 6022–6034. https://www.aclweb.org/anthology/2020.acl-main.536
27. Wenzek G, Lachaux M-A, Conneau A, Chaudhary V, Guzman F, Joulin A, Grave E (2019) CCNET: extracting high quality monolingual datasets from Web crawl data. arXiv:1911.00359
28. Wu D (1994) Aligning a parallel English-Chinese corpus statistically with lexical criteria. In: Proceedings of the 32nd annual meeting on Association for Computational Linguistics (ACL'94), Stroudsburg, PA, USA. Association for Computational Linguistics, pp 80–87
29. Wu S, Dredze M (2019) Beto, Bentz, Becas: the surprising cross-lingual effectiveness of BERT. arXiv:1904.09077
30. Wu S, Dredze M (2020) Are all languages created equal in multilingual BERT? In: Proceedings of the 5th workshop on representation learning for NLP, Online, July 2020. Association for Computational Linguistics, pp 120–130. https://doi.org/10.18653/v1/2020.repl4nlp-1.16. https://aclanthology.org/2020.repl4nlp-1.16
31. Xiao R (ed) (2020) Using corpora in contrastive and translation studies. Cambridge Scholars Publishing
32. Xu H, Koehn P (2021) Cross-lingual BERT contextual embedding space mapping with isotropic and isometric conditions. arXiv:2107.09186

3 Building Comparable Corpora

In a parallel corpus we know which document is a translation of what by design. If the link between documents in different languages is not known, it needs to be established. In this chapter we will discuss methods for measuring document similarity across languages and how to evaluate the results. Then, we will proceed to discussing methods for building comparable corpora of different degrees of comparability and for different tasks. We will start with the easiest case here, when there is *natural annotation*, i.e. comparable documents are explicitly linked by design. Usually this leads to strongly comparable corpora. The next step concerns getting the low-hanging fruit by looking for documents which are not linked, but which can be easily linked across languages. Finally, we can present methods for aligning large collections of unrelated documents from weakly comparable corpora in the hope of finding some documents which are similar across the languages.

The document alignment task can be formalised as the maximum-weighted bipartite matching (MWBM) problem [1]. We start with a weighted bipartite graph $G = (V_1 \cup V_2, E)$ with documents V_1 in Language$_1$ and documents V_2 in Language$_2$ as the nodes and E as edges between V_1 and V_2 with the corresponding edge weights $c_{i,j}$. The task is to find a matching $M \subset E$ which links closely related documents, for example, for the maximal values of the edge weights $c_{i,j}$. The challenge for this task is to define a cross-lingual document similarity measure to compute $c_{i,j}$ and to tune its threshold to select M. Given that E links every V_1 to every V_2, the size of M is a tiny proportion of E even when each V_1 has its candidate among V_2, which is not always the case.

A related problem concerns the document boundaries. It often happens that for two related documents only a portion of one document offers a reasonably close translation to its counter-part in another language. For example, a Wikipedia page in English is often considerably longer than its equivalent in a lesser resourced language. For the purposes of

S. Sharoff et al., *Building and Using Comparable Corpora for Multilingual Natural Language Processing*, Synthesis Lectures on Human Language Technologies,
https://doi.org/10.1007/978-3-031-31384-4_3

this chapter the documents are treated as single entities, while Chap. 4 below addreses the task of finding translatable subsets within pages.

An attempt at aligning documents between different languages is useful, because by using this information we can get more precise *local* information about the nature of the mapping between linguistic structures in different languages. However, this is a task with diminishing returns, as out corpora are getting bigger and less related, the vast majority of documents in Language_1 do not have translations in Language_2. Therefore, if we aim at achieving better coverage, we can also drop the requirement of document alignment, while still being able to detect some general patterns which link the languages at the level of linguistic constructions. Through this we can still derive useful resources, such as multilingual dictionaries or cross-lingual prediction models, see specific examples in Sects. 5.4 concerning the dictionaries and 7.1 concerning zero-shot transfer.

3.1 Measures for Document Similarity Across Languages

The similarity between documents in the two languages for building a comparable corpus is usually estimated following the Vector Space Model (VSM) representation. If documents in two languages are described via a shared VSM space, i.e., documents i and j in two languages are represented as vectors E_i and F_j, their similarity is assessed through a similarity function between these two vectors, for example, the cosine similarity:

$$\cos(E_i, F_j) = \frac{E_i \cdot F_j}{\|E_i\| \|F_j\|}$$

The similarity between the vectors can be also measured in different ways, for example, via their Euclidean or Manhattan distances [2]. It is also possible to measure their similarity by using the correlation scores between the values of the respective vectors, often leading to better cross-lingual similarity predictions in comparison to more traditional distance measures [3].

The differences between approaches to the document alignment task are largely in using different methods for building the shared VSM space and in using different similarity functions. The simplest case is to extract keywords with their weights for E_i and F_j^o (o for original). The weights can be, for example, the Pointwise Mutual Information values for the frequency of the keywords in a document vs their frequency in the corpus, see Equation (2.1). Other methods for keyword extraction can also be used, such as TF-IDF [4]. Once the keywords have been extracted, the keywords of the original vectors F_j^o can be translated into the keyword set of E_i using a dictionary or Machine Translation. Instead of simply mapping the values of weights for listed keyword translations via a dictionary function, it is also possible to use a translation matrix **Tr** [5] which represents the probabilities of translating w_f into w_e as:

$$F_j = \mathbf{Tr} \times F_j^o$$

The rise of neural methods led to the use of word embedding spaces which can be shared across languages. These word-level spaces can be used to generate document-level vectors of a fixed length for building document-level vector spaces, in which similar documents are still positioned close to each other. For example, this can be achieved by learning a document matrix along with the word embedding matrix [6]. In the end, document-level embedding spaces can be also aligned across different languages [7].

3.2 Evaluation Methods and Datasets

The purpose of shared tasks is to establish a gold standard for comparing different methods, in this case for detecting which documents are likely to be similar. In this section we describe the corpora and tools for evaluation as used in a BUCC'15 shared task [11]. A successor Bilingual Document Alignment Shared Task similar in design has been tried at WMT-16 [12].

The original dataset for the BUCC'15 shared task is based on static Wikipedia dumps from 2011. In preparing the dataset we relied on iWiki links between corresponding articles with the dataset split into the training set with correct links provided to the participants and the test set, a set of articles with known language codes, but no information about links. The language directions were from Chinese, French, German, Russian and Turkish aligned to English. The choice of languages ensured variation with respect to:

their writing systems Cyrillic is used for Russian and logographic writing for Chinese. This helps in estimating the reliance of some algorithms on the influence of identical strings, such as place names, as they have to be transliterated for those languages:
Париж ('Paris', transliterated as *Parizh* in Russian) or
巴黎 ('Paris' pinyin *Bali* in Chinese)

tokenisation many methods rely on identification of individual words, which can be challenging for some languages, for example, clitics are used in French, compound words in German, there are no orthographic word boundaries in Chinese;

morphology the aim is to estimate the influence of the number of word forms expressing the same meaning, as having more forms leads to a sparser representation for mapping words with similar meanings across languages. In our dataset, Chinese is an isolating language, which does not have separate forms of the same word, Russian has richer morphology, where adjectives, for example, can have different endings to reflect the values with respect to two numbers, three genders and six cases, while Turkish is an agglutinative language which builds new forms by gluing relevant prefixes and suffixes to the root to produce an even larger number of forms, more than 100 for a verb.

Wikipedia is a widely used resource, which we improved by the following filtering rules:

- iWiki links had to be bidirectional, an article in Language_1 linked to an article in Language_2, but also vice versa (occasionally a Wikipedia page in Language_1 are aligned to a section of a relevant Wikipedia page in Language_2).
- Articles had to be of roughly similar size in order to avoid outlier document pairs with considerable differences in their length, since they are less likely to be a source of useful information for translation purposes.

Table 3.1 shows the distribution of text lenghts and the number of corresponding articles selected for the shared task dataset. Some articles are much shorter than their English counterparts. Less frequently this happens in the opposite direction, while the median ratio of article lengths (Language_X/English) is always less than one. Usually articles which differ in their length are not good candidates for comparable corpora, as they include much more information in one language in comparison to the other. We took only those articles for which the ratio of lengths was within the inter-quartile range for this language. This left us with 50% of article pairs with respect to the original list, which are all reasonably comparable since they belong to the same narrow topic.

For converting Web pages to a corpus we removed the page title, boilerplate (such as *From Wikipedia..*), images and tables with their captions, the table of contents (shaded areas in Fig. 3.1 indicate removed text). We also removed the first sentence of every page, since the corpus is intended to be the benchmark for linking arbitrary Web pages, while the very genre of Wikipedia assumes that the first sentence defines the topic of the article and by its own the first sentence in Wikipedia articles gives a clue for cross-lingual similarity.

Table 3.1 Ratios of lengths to English

	Min.	1st Qu	Median	Mean	3rd Qu	Max.	Selected pairs
De	0.010	0.420	0.790	1.244	1.370	206.000	294990
Fr	0.000	0.370	0.740	1.194	1.260	255.800	229591
Ru	0.010	0.300	0.620	0.987	1.070	108.600	159810
Tr	0.000	0.140	0.350	0.616	0.760	46.730	32614
Zh	0.010	0.280	0.610	1.010	1.090	111.500	42944

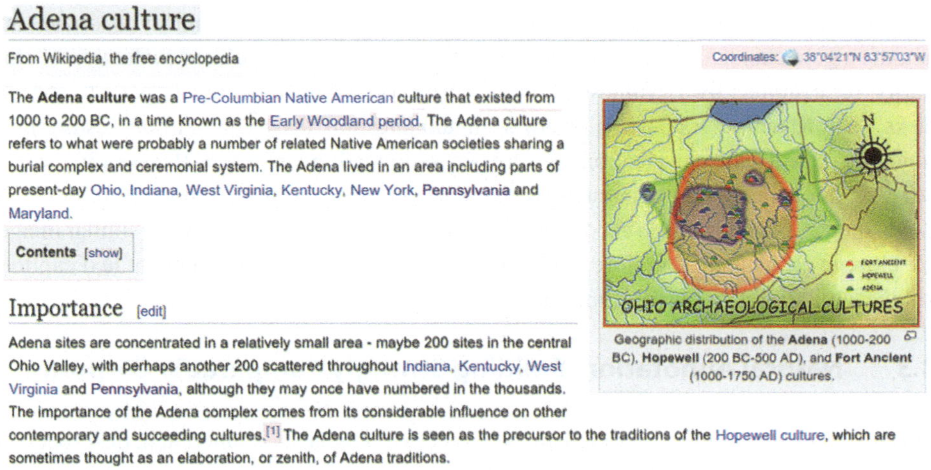

Adena culture

From Wikipedia, the free encyclopedia

Coordinates: 38°04′21″N 83°57′03″W

The **Adena culture** was a Pre-Columbian Native American culture that existed from 1000 to 200 BC, in a time known as the Early Woodland period. The Adena culture refers to what were probably a number of related Native American societies sharing a burial complex and ceremonial system. The Adena lived in an area including parts of present-day Ohio, Indiana, West Virginia, Kentucky, New York, Pennsylvania and Maryland.

Contents [show]

Importance [edit]

Adena sites are concentrated in a relatively small area - maybe 200 sites in the central Ohio Valley, with perhaps another 200 scattered throughout Indiana, Kentucky, West Virginia and Pennsylvania, although they may once have numbered in the thousands. The importance of the Adena complex comes from its considerable influence on other contemporary and succeeding cultures.[1] The Adena culture is seen as the precursor to the traditions of the Hopewell culture, which are sometimes thought as an elaboration, or zenith, of Adena traditions.

Geographic distribution of the **Adena** (1000-200 BC), **Hopewell** (200 BC-500 AD), and **Fort Ancient** (1000-1750 AD) cultures.

Fig. 3.1 Example of Wiki-page cleaning for comparable corpora

In the dataset for each source page there exists exactly one correct linked page in the gold standard. Systems were required to return a ranked list of hypotheses in which the correct target page should be ranked as high as possible.

We used the standard Information Retrieval evaluation script from TREC[1] and from its vast array of evaluation measures we selected those relevant for our evaluation goals:

Success@1 Proportion of source articles for which the correct target article is in the top position, this is the label from the TREC tool, in other NLP methods this is usually called Precision@1;

Success@5 Proportion of source articles for which the correct target article is among the top 5 positions;

Mean Reciprocal Rank If the correct target article is at position N, $1/N$ is given to this source article. This is then averaged over the set of source articles.

Difference in measures:

- Success@1 (Precision@1) is the strictest measure;
- Mean reciprocal rank (MRR) adds decreasing fractions of one when the correct article is found below the first in the ranking, so that this leads to a higher average score than success@1;
- Success@5 also takes into account articles beyond the first, but only until the fifth, which means that a system which ranks correct articles beyond the fifth position success@5 $<$ MRR, while a system which ranks correct articles within top five success@5 $>$ MRR;

[1] http://trec.nist.gov/trec_eval/.

This is an intrinsic evaluation procedure, which compares the similarity of documents against a gold standard of cross-lingual links. One can also run extrinsic evaluation of similarity, for example, by usefulness of the retrieved cross-lingual document pairs to specific tasks, such Machine Translation or cross-lingual question answering. An interesting multimodal application for evaluation concerns cross-lingual image retrieval tasks [36], which starts with a collection of documents with associated images, such as image captions from social media, and proceeds to encoding images into embedding spaces, so that similar images are close to each other. Therefore, one can estimate the quality of alignment of text captions via the similarity of respective images.

3.3 Natural Annotation: Building Strongly Comparable Corpora

In many cases some kind of parallelism in texts across languages is inherent in the process of their production, thus leading to *natural* annotation of resources similar in several languages. One case of natural parallelism is the use of translations, as translations are produced for purposes unrelated to the task of building corpus resources and they are produced in substantial quantities, for example, from instruction manuals, subtitles [13] or the institutional repositories, such as the European Parliament [14] or the United Nations [15].

As noted in the introduction, there are limits to such sources of natural annotation for faithfully translated texts with respect to domains, genres and the presence of translationese. However, we can consider other cases of natural annotation, in which texts in different languages are explicitly linked and which provide more data in comparison to exact translations. Some of the widely used sources of this kind are Wikipedia articles and patents, in which explicit identifiers link texts which are not necessarily precise translations.

The page titles in Wikipedia which are linked via iWiki links provide reliable translations. Merely the titles themselves can be used as a source of words for a seed dictionary, especially in cases when bilingual resources are scarce, see examples in Table 3.2. In addition to this "parallel" corpus, each individual page, for example, *Budapeszt* for Polish and *Budapest* for English, can be treated as good document-level candidates for retrieving translated sentences. Even though Wikipedia pages are rarely exact translations for each other because of the number of Wikipedia editors in each individual language, often the first version of each page is produced by translating an existing page from a more dominant language (for

Table 3.2 Alignments from Wikipedia for titles and words

Polish	Russian	English
Budapeszt	Будапешт	Budapest
Kapral	Капрал	Corporal
Z życia marionetek	Из жизни марионеток	From the Life of the Marionettes
Wskaźnik jakości życia	Индекс качества жизни	Quality-of-life index

example, English), so the Wikipedia pages have been used for mining similar sentences early on [16]. Experiments with their large scale mining demonstrated the limits of detection of similar sentences using very simple mechanisms, such as identical words [17]. Application of modern neural methods, which rely on the similarity of embeddings rather than on individual words, results in much better precision and recall rates [75]. For more information on extracting parallel sentences from such resources, see Chap. 4.

Patents are fairly parallel resources, even though there is no requirement for having exactly the same submission to different national patent agencies, as they follow different templates and practices. Instead alignment of patents is based on the presence of the document identifiers as well as the domain codes [18].

3.4 Low-Hanging Fruit: Building Weakly Comparable Corpora

The international news items are inherently similar, as they report similar events even when they are produced by unrelated news agencies. When they are produced by the same agency, they can contain a good portion of translations of each other even when they have language-specific focus points. For example, the BBC news are produced for different audiences, e.g., British English and Romanian, with a large amount of Romanian news translated from the original English source [19].

Similarly the Reuters News Agency released a multilingual corpus of newswires for nine languages [20], which covers the same time range and the same set of categories for all of the languages. This has been used as a source of experiments for a number of cross-lingual document classification (CLDC) models, for example, [21, 22]. At the same time, a real corpus of this kind creates an unbalanced distribution of labels across the languages. For example, as noted in [23], the CCAT label (Corporate news) in the RCV2 test set covers 1.5% of texts in English and 21.4% of texts in French, while the MCAT label (Martket news) has a very different distribution of 46.6% vs 12.8% respectively, so the more successful approaches dealing with the CLDC task primarily focus on tricks for dealing with this imbalance rather than on genuinely multilingual approaches. Therefore, Schwenk and Li created a standardised subcorpus for RCV2 with uniform coverage of class labels across the languages, as well as the standard train, development and test sets.

Even when unrelated news agencies report on the same topic, their news streams can be aligned through the use of simple heuristics, such as Named Entities (names of people, locations and organisations) and dates, as used in the JRC news monitoring service [24]. The resulting clusters can be considered as strongly comparable texts.

In addition to direct crawling, a different way for obtaining weakly comparable corpora is by making queries to search engines. This relies on crawling already performed by the search engines. If the queries are reasonably parallel, see examples in Table 3.3, the resulting corpora are likely to lead to reasonably similar texts, see analysis on collecting comparable corpora in specific domains [5].

Table 3.3 Parallel queries for comparable texts resources

English	Chinese	Russian
fossil fuel	化石燃料	ископаемоетопливо
power station	发电厂	электростанция
hydroelectricity	水力发电	гидроэнергетика
photovoltaics	太阳能光伏	фотоэлектричество

3.5 Large Scale Document Alignment

3.5.1 Structural Similarity

While Wikipedia and patents have natural annotations because of the very process of their development, many web pages contain texts with their approximate translations without explicit iWiki-kind links or without making queries which are known to be the translations of each other.

The first approach of this kind was PTMiner from Montreal, [28].

However, some links between possible translations can be detected via the similarity in their URLs. For example, two URL links http://xx.zz/index-en.html and http://xx.zz/index-fr.html are likely to refer to two noisy parallel texts (following the scale from Sect. 1.2), which can differ in their content because of their intended audiences or the amount of attention paid to the process of translation by the website owner. Nevertheless, they present good candidates for alignment, which can be detected using simple rules.

If the the two candidate documents are formatted as HTML pages, they can contain both lexical and structural similarities, for example:

<HTML>	<HTML>
<TITLE>Emergency Exit</TITLE>	<TITLE>Sortie de Secours</TITLE>
<BODY>	<BODY>
<H1>Emergency Exit</H1>	Si vous êtes assis à
If seated at an exit and	côté d'une ...

Nevertheless some structural and lexical elements might be missing, for example, the H1 tag and its content in this example. This sequence can be aligned through matching HTML tags while detecting and ignoring missing segments, for example:

1	[START:HTML]	[START:HTML]
2	[START:TITLE]	[START:TITLE]
3	[Chunk:13]	[Chunk:15]
4	[END:TITLE]	[END:TITLE]
5	[START:BODY]	[START:BODY]
6	[START:H1]	
7	[Chunk:13]	
8	[END:H1]	
9	[Chunk:112]	[Chunk:122]

This was the starting step for STRAND, one of the first experiments for extracting noisy parallel data from the Web [29]. It was based on three steps:

1. locating parent pages referring to collections of parallel texts using Web search for pages which can contain noisy parallel texts,
2. generation of document candidates using rule-based URL matching techniques from similarities of their parallel URL patterns,
3. structural filtering with four variables for assessing the quality of their alignments:

- dp The difference percentage, indicating nonshared material (i.e., alignment tokens that are in one aligned file but not the other).
- n The number of aligned nonmarkup text chunks of unequal length.
- r The correlation of lengths of the aligned nonmarkup chunks.
- p The significance level of the correlation r.

In the initial experiments the thresholds for filtering have been set manually, such as choosing a link if $dp < 0.2$ and $p < 0.05$. Later on, the thresholds have been optimised by training a decision tree algorithm on a small training set. Extensions to content-based matching involve using a bilingual lexicon, which adds a score of the proportion of word translation pairs available in the two alignment candidate documents.

An expansion of the STRAND algorithm was applied to Common Crawl [30] as a simple way of obtaining millions of parallel page candidates. A further extension of this approach led to the ParaCrawl corpus [32] which consists of noisy parallel sentences, see more information in Chap. 4.

The corresponding pipeline for crawling and aligning Web documents with similar structure is available in open source [33].

3.5.2 Lexical Similarity

The methods presented above used pre-defined structural conditions to find texts which are likely to be comparable by combining the URL filters and segmentation alignment. A more challenging approach involves aligning arbitrary texts under the assumption that loose translations and very similar documents can be found by seiving through millions or billions of unrelated documents, only some of which might be comparable.

A direct approach of comparing all pairs in this collection is not feasible because of the quadractic computational costs. A simple approach to circumvent this issue is by taking frequent n-grams in one language and translating them using machine translation. The n-grams can be selected via TF-IDF or similar keyness measures to create signatures for each document in both languages. Machine-translated n-grams save the amount of translation which would be needed for translating the pages. This also creates a pivot-language space, which allows application of very efficient methods for detecting near-duplicates [37], which are good alignment candidates. The Google Translate team used it to mine very large corpora for their translation service [39]. A modern variation of this approach [42] compares documents by projecting documents of a web domain to a shared multilingual space by using a word-based translation model and using TF-IDF values to create document vectors.

A very simple approach, which is also surprisingly efficient, involves the use of Lucene by making monolingual queries for identical elements, such named entities or URLs, for which no translation is required [17], or by making bilingual queries, which involve translation of the original query using a bilingual lexicon [43]. Purely monololingual approach reaches the Precision@1 and Precision@5 scores of 76.6% and 92.6% on the WMT16 shared task dataset. With sufficient filters for the bilingual lexicon (as derived from an SMT engine), the scores on the same dataset improve to 83.3% and 96.2%. At the same time, the approaches which involve identical elements are limited to language pairs of identical orthography, such as the English-French pair tested in the WMT16 dataset, while a good coverage dictionary of word-form translations needed for the bilingual queries itself sufficient amount of resources, see also Bilingual Lexicon Induction methods in Chap. 5.

The number of comparisons between the alignment candidates in a dataset rises quadratically with the corpus size, thus making many methods less suitable for Web-sized text collections. A simple method for prefiltering alignment candidates [44] has been suggested in order to select in-domain corpora by computing cross-entropy for a word sequence as:

$$H_{LM}(W) = -\frac{1}{n}\sum_{i=1}^{n} \log(P_{LM}(w_i|w_1, ..., w_{i-1}))$$

where P_{LM} is the probability of a LM for the word sequence W (normally a sentence). Comparison of sequences can be estimated via the difference between $H_{LM_{I,b}}(W_O) - H_{LM_{O,b}}(W_O)$ where $b \in e, f$ refers to the sides of the in-domain vs out-of-domain corpora. It has been shown that using cross entropy can lead to substantial gains, while decreasing computation time by a factor of 10 in comparison to fine-tuning BERT for the document classification task.

As machine translation is becoming more popular on Web pages, alignment of Web pages is not sufficient for the purpose of obtaining translations naturally produced by humans,

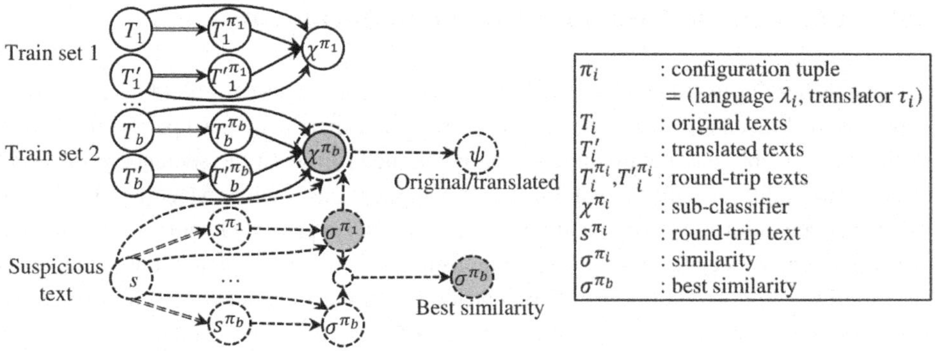

Fig. 3.2 Round-trip translation for detecting MT text

because many parallel texts will be actually output by MT. This leads to the need of employing methods for filtering MT output from comparable corpora. In particular, it has been shown that the use of ngrams is sufficient to detect and reject MT texts, as rule-based MT tends to have awkward lexical choices, while statistical MT often suffers from re-ordering problems [46].

Since the publication of [46], modern NMT engines started producing considerably more fluent texts, which contain fewer awkward lexical choices and fewer re-ordering problems. This means that the use of n-grams is becoming less efficient, while the presence of MT-generated texts presents a less pressing problem. Nevertheless, it is still advisable to filter the machine-translated texts out from comparable corpora, as their presence can pollute comparable resources by preventing improvements in MT quality through simply re-using existing translation choices rather than contributing new texts. Neural pipelines for detecting machine-produced texts involve fine-tuning classifiers for making the prediction task on whether a suspicious text has been produced by MT or not [47]. Another approach involves adding round-trip translations via several languages to increase variation translated texts [50]. Figure 3.2 illustrates the principle of building classifiers by generating two training sets using a number of source texts (T) with their available human translations (T') and a number of MT engines which can produce variations of the source texts via round-trip translations ($T^{\pi}_{i,j}$), i.e., translations into language λ_i and back into the original language using an MT engine τ_j. This gives the possibility to train several sub-classifiers (for a combination of available languages and translation engines) to distinguish between naturally produced translated texts and texts which under-gone round-trip translation. In [50] the sub-classifiers have been made by simply fine-tuning BERT. To predict whether a suspicious text s is a piece of original production or it is a machine-translated text, we first create its round-trip translations $\sigma^{\pi_{i,j}}$ for the λ_i, τ_j setting of each sub-classifier and then we select the best sub-classifier to make the predition by the minimum value of the Levenshtein distance between s and $s^{\pi_{i,j}}$.

3.6 Comparable Corpora of Unrelated Documents

The approaches discussed in Sects. 3.3–3.5 assume that the purpose for building comparable corpora is to align candidates at the document level. This results in quasi parallel resources, in which reasonably parallel documents can be aligned further at the sentence level which makes them useful for multilingual downstream applications, such as Machine Translation. However, many multilingual applications can avoid fairly expensive document- or sentence-level alignment, especially when the documents in corpora are completely unrelated, hence there are fewer options for finding document-level alignment candidates, while we can still find ways of utilising them for the purposes of building multilingual models.

As comparable corpora of unrelated documents can be large and representative for reflecting many phenomena in each individual language, they can be very useful to establish the similarity between different languages, for example, for comparing the frequency lists across languages, see the example of language teaching in Sect. 7.2, or for training word embedding models, which can establish reasonable similarity of their representation space with simple linear transformations by using small seed dictionaries [52] or even in an unsupervised way by utilising similar probability distributions [53], see also Sect. 5.4.

A typical example of how unrelated comparable corpora can be built from the Web concerns comparable Web crawls, when large corpora for several languages are collected from the Web independently, while still using very similar pipelines [55, 56]. Usually, Web crawling starts from a random set of URLs for each language, proceeds with detecting all URLs mentioned in those pages to enrich the pool of available Web pages and repeats the procedure of URL detection with the updated pool [61].

A good example of comparable corpora from unrelated document collections of very large sizes is CC-Net [62] which is derived from Common Crawl, a large-scale Web crawl which has been deduplicated, filtered with respect to the languages of its Web pages and cleaned by applying the respective language models to remove non-text and machine generated text, see the counts for the biggest languages in CC-Net in Fig. 3.3. By the nature of its production, the proportion of parallel pages in collection is relatively small. Nevertheless,

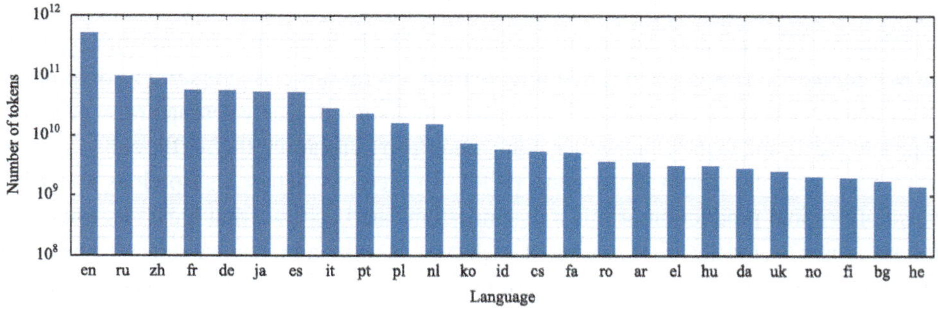

Fig. 3.3 Number of tokens per language in CC-Net in logarithmic scale

this set of comparable corpora has been used in building XLM-Roberta [68], a multilingual pre-trained transformer which has been successfully applied to a number of multilingual transfer tasks, see Sect. 7.1.

The success of models based on unrelated comparable collections stems from the assumption that such crawls for different languages lead to somehow similar pages. This leads to a question of measuring similarity of unrelated corpora. The quality of comparable corpora with linked pages can be assessed by the degree of alignment of their individual documents. Since this is not possible with unrelated comparable corpora, we need methods to estimate the similarity of their composition as a whole. This task can be approached in several ways, for example, by comparing the frequencies of their words or n-grams via translations, by comparing unsupervised topic models or by running supervised classification [69].

Generation of topic models is based on Latent Dirichlet Allocation (LDA), which estimates the distribution of probabilities of keywords belonging to different topics as well as the proportions of documents over the same set of topics. This is an unsupervised procedure, in which unknown distributions of latent variables are derived in repeated approximations [70]. The probabilities of keywords over documents are observed, while the probabilities of keywords over topics and the probabilities of topics over documents are derived as latent variables. For each topic $\vec{\beta}_k$ $(k = 1 \ldots K)$ the task is to obtain the distribution of the probabilities of keywords over topics:

	NBA	Players	Season	Steam	…	Xbox
$\vec{\beta}_1$	0.007	0.009	0.015	0.000	…	0.000
$\vec{\beta}_2$	0.000	0.010	0.002	0.010	…	0.011

The LDA model also estimates the distribution of the degree its documents $\vec{\theta}_d$ $(d = 1 \ldots T)$ belong to topics:

$$\vec{\theta}_1 : (0.783\beta_1, 0.002\beta_2, ...0.122\beta_K)$$

$$\vec{\theta}_2 : (0.002\beta_1, 0.550\beta_2, ...0.213\beta_K)$$

Instead of estimating the probabilities of specific individual keywords, for a very large vocabulary of keywords coming from large unrelated corpora it is often beneficial consider neural Embedding Topic Models as they match embeddings of keywords to embeddings of topics [71].

In the end, topic modelling assigns each document to a vector of topic probabilities. An estimation for the relative proportion of topics in a corpus can be provided by summing up the vectors of topics over all documents. Table 3.4 lists the keywords assigned to the most used topics in CC-Net for English, French and Russian. A good proportion of the most used topics are shared across the languages in the case of the CC-Net family, for example, medicine (Topic 69 for English, Topic 36 for French and Topic 67 for Russian) or entertainment (Topic 22 for English, Topic 22 for French and Topic 59 for Russian).

Table 3.4 Topics detected in CC-Net for English, French and Russian

ccnet-en
Topic 45 president, government, trump, party, political, election, state, united, obama, security, states, israel, country
Topic 69 cancer, disease, blood, body, patients, treatment, health, pain, heart, symptoms, risk, skin, surgery
Topic 80 team, game, season, players, games, league, sports, football, play, win, teams, player, year
Topic 22 film, movie, series, films, story, movies, show, tv, star, episode, character, season, comedy
Topic 90 park, city, hotel, area, located, island, beach, beautiful, visit, town, river, travel, visitors
Topic 95 court, law, legal, state, act, case, states, federal, attorney, government, rights, supreme, section
Topic 88 music, album, band, song, songs, rock, musical, sound, dance, guitar, concert, show, released
Topic 71 million, market, percent, year, billion, price, growth, rate, stock, prices, average, increase, company
Topic 48 energy, water, car, power, gas, air, system, solar, oil, electric, cars, vehicle, vehicles
Topic 55 god, church, jesus, faith, life, christ, lord, christian, bible, holy, love, spirit, prayer
Topic 13 internet, mobile, users, google, computer, game, software, web, data, online, devices, digital, system
Topic 58 business, company, industry, market, marketing, services, technology, customers, product, sales
Topic 67 police, crime, court, criminal, case, law, prison, accident, charges, death, man, county, injury
Topic 28 atudents, school, education, college, learning, program, teaching, academic, programs, university
Topic 16 love, back, life, feel, things, thought, family, friend, friends, good, man, told
ccnet-fr
Topic 36 patients, symptômes, maladie, peau, cellules, effets, cancer, dents, mèdicaments, troubles, douleur, diabète
Topic 21 jésus, foi, christ, saint, seigneur, parole, pape, mort, catholique, église, qu, chrétiens, père, fils, bible, prière
Topic 63 client, personnelles, commande, www, fr, com, paiement, réserve, utilisation, présentes, responsable
Topic 17 croissance, milliards, taux, banque, hausse, dollars, baisse, crise, banques, fonds, marchés, dette, financiers
Topic 42 château, église, roi, saint, chapelle, louis, ii, pierre, iii, musée, ier, village, rome, patrimoine, époque
Topic 61 roman, personnage, récit, auteur, livres, femme, littérature, lecture, romans, scène, mort, lecteur, série
Topic 51 russie, guerre, syrie, trump, israël, forces, gouvernement, palestiniens, militaire, turquie, américain, paix
Topic 45 mouvement, peuple, liberté, démocratie, révolution, lutte, capitalisme, guerre, crise, discours, peuples, parti
Topic 90 cinéma, films, the, série, album, scène, réalisateur, festival, saison, star, succès, of, chanson, quot, rock,
Topic 44 pensée, physique, science, théorie, conscience, philosophie, réalité, lumière, humain, connaissance, choses
Topic 29 logement, taux, montant, frais, assurance, crédit, immobilier, prêt, retraite, revenu, bénéficier, régime, salariés
Topic 53 électrique, véhicule, voiture, moteur, chaleur, chauffage, bois, vitesse, électriques, installation, automobile
Topic 75 cour, juge, justice, tribunal, procédure, décision, avocat, judiciaire, juridique, parties, responsabilité, civil
Topic 52 google, web, utilisateurs, mobile, facebook, numérique, logiciel, apple, vidéo, téléphone, applications, ordinateur

(continued)

Table 3.4 (continued)

ccnet-ru
Topic 67 заболевания, крови, лечение, болезни, симптомы, препарата, печени, боль, организм, препарат
Topic 56 воды, материал, конструкции, устройство, воздуха, оборудование, эксплуатации, установки
Topic 21 турист, отдых, реки, территории, воды, парк, море, расположен, остров, горы, музей
Topic 43 федерации, республики, органов, государственной, местного, федерального, закон, субьект
Topic 84 чай, масло, продукт, питания, растения, организма, плоды, веществ, вкус, масла
Topic 42 суд, дела, федерации, прав, ответственности, лиц, решения, закона, кодекс, срок, действия
Topic 59 фильм, онлайн, кино, роль, альбом, игра, сериал, актер, шоу, видео, режиссер, актриса
Topic 83 империи, век, здание, их, название, церковь, xviii, территории, земли, культуры, король
Topic 33 участие, конкурс, культуры, района, молодежи, мероприятие, акции, детского, участники
Topic 99 материал, дизайн, изделия, вид, пвет, интерьер, конструкции, двери, мебель, выбор, дерева
Topic 26 чемпионат, спорта, сборной, команда, матч, кубок, динамо, лиги, турнира, европы, футболу, тренер
Topic 20 личности, общества, культуры, поведения, философии, отношений, понятие, социальной, психологии
Topic 77 обучение, образование, школы, образовательных, подготовки, программа, язык, дошкольного
Topic 97 армии, великой, отечественной, советского, 1941, войска, союза, 1945, военного, красной, полка

More formally, the degree of similarity between the topics across languages can be assessed via the Jensen-Shannon divergence [72] as follows:

$$D_{JS}(\vec{\beta}_1, \vec{\beta}_2) = \frac{D_{KL}(\vec{\beta}_1, \vec{B}) + D_{KL}(\vec{\beta}_2, \vec{B})}{2}$$

where $\mathrm{D_{KL}}(\vec{x}, \vec{y}) = \sum x_i \ln(x_i/y_i)$ is the Kullback-Leibler divergence and $\mathbf{B} = \frac{\vec{\beta}_1 + \vec{\beta}_2}{2}$ provided that we have a translation dictionary for the keywords.

Topic modelling does not require supervision data, so it can be applied to assessing the similarity of arbitrary pairs of unrelated comparable corpora. However, this requires the need to match topics produced in different ways in different languages, while the coverage of translation dictionaries is not always good and translation equivalents often depend on their context of use, so it is difficult to rely on translations of isolated keywords, such as *park* for Topic 90 or *team* in Topic 80 in English in Table 3.4. Another way to assess variation between unrelated corpora is by using supervised classifiers with the same set of labels across the languages. Table 3.5 lists the genres predicted for the CC-Net corpora for English, French

Table 3.5 Distribution of genre categories in pretrained corpora

OpenWebText		CC-Net-en		CC-Net-fr		CC-Net-ru	
42.01%	Argument	29.60%	Promotion	23.20%	Argument	28.33%	News
14.91%	Personal	20.01%	Argument	19.03%	News	27.26%	Promotion
4.96%	News	14.52%	news	18.54%	promotion	19.46%	Information
3.36%	Instruction	12.49%	Information	17.77%	Information	8.84%	Argument
2.75%	Review	9.64%	Personal	6.91%	Review	6.53%	Legal
1.96%	Promotion	6.03%	Instruction	4.26%	Legal	3.23%	Academic
1.85%	Academic	4.25%	Review	4.17%	Instruction	2.81%	Instruction
0.87%	Information	2.42%	Legal	4.05%	Personal	2.39%	Review
0.85%	Legal	0.67%	Academic	1.51%	Academic	1.08%	Personal
0.03%	Fiction	0.37%	Fiction	0.57%	Fiction	0.08%	Fiction

and Russian, as well as the genres predicted for a monolingual English corpus OpenWebText, which was used for pre-training GPT-2, a popular monolingual language model [73], and which was built from a pipeline different from crawling, namely by extracting Web pages from all URLs upvoted 3 or more times on the Reddit website. While the composition of CC-Net corpora is roughly similar across the three languages with the most common kinds of texts concerning news, commercial promotion and reference information (such as Wikipedia), the different pipelines used in OpenWebText and CC-Net lead to very different corpora. The degree of their differences can also be measured, either through measuring correlation between the coverage of the predicted categories (corpora with better correlation between the predicted categories are closer to each other) or by treating the predicted categories as a vector and measuring the cosine distance between the vectors [74].

These evaluation approaches are still intrinsic. Extrinsic evaluation can be performed via using the corpora in specific tasks, as discussed further in Chap. 7.

References

1. Melamed ID (2000) Models of translation equivalence among words. Comput Linguist 26(2):221–250. https://aclanthology.org/J00-2004
2. Manning C, Schütze H (1999) Foundations of statistical natural language processing. MIT Press, Cambridge, MA
3. Forsyth R, Sharoff S (2014) Document dissimilarity within and across languages: a benchmarking study. Literar Linguist Comput 29:6–22
4. Schütze H, Manning CD, Raghavan P (2008) Introduction to information retrieval, vol 39. Cambridge University Press Cambridge
5. Sharoff S (2013) Measuring the distance between comparable corpora between languages. In: Sharoff S, Rapp R, Zweigenbaum P, Fung P (eds), BUCC: building and using comparable corpora. Springer, Berlin/New York, pp 113–129

6. Le Q, Mikolov T (2014) Distributed representations of sentences and documents. In: International conference on machine learning, pp 1188–1196. PMLR
7. Yarmohammadi M, Ma X, Hisamoto S, Rahman M, Wang Y, Xu H, Povey D, Koehn P, Duh K (2019) Robust document representations for cross-lingual information retrieval in low-resource settings. In: Proceedings of machine translation summit XVII: research track, Dublin, Ireland, Aug 2019. European Association for Machine Translation, pp 12–20. https://aclanthology.org/W19-6602
8. Zanettin F (1998) Bilingual comparable corpora and the training of translators. Meta: J des Traducteurs 43:12. https://doi.org/10.7202/004638ar
9. Zhao B, Vogel S (2002) Adaptive parallel sentences mining from web bilingual news collection. In: Proceedings of the 2002 IEEE international conference on data mining, pp 745–748. https://doi.org/10.1109/ICDM.2002.1184044
10. Zhao W, Eger S, Bjerva J, Augenstein I (2020) Inducing language-agnostic multilingual representations. arXiv:2008.09112
11. Sharoff S, Zweigenbaum P, Rapp R (2015) BUCC shared task: cross-language document similarity. In: Proceedings of the eighth workshop on building and using comparable corpora, Beijing, China, July 2015. Association for Computational Linguistics, pp 74–78
12. Buck C, Koehn P (2016) Findings of the WMT 2016 bilingual document alignment shared task. In: Proceedings of the first conference on machine translation: volume 2, Shared Task Papers, Berlin, Germany, Aug 2016. Association for Computational Linguistics, pp 554–563. https://doi.org/10.18653/v1/W16-2347. https://aclanthology.org/W16-2347
13. Tiedemann J (2012) Parallel data, tools and interfaces in OPUS. In: Proceedings of the eighth international conference on language resources and evaluation (LREC'12), Istanbul, Turkey, May 2012. European Language Resources Association (ELRA), pp 2214–2218. http://www.lrec-conf.org/proceedings/lrec2012/pdf/463_Paper.pdf
14. Koehn P (2005) Europarl: a parallel corpus for statistical machine translation. In: Proceedings of the MT Summit X
15. Ziemski M, Junczys-Dowmunt M, Pouliquen B (2016) The United Nations parallel corpus v1.0. In: Proceedings of the tenth international conference on language resources and evaluation (LREC'16), Portorož, Slovenia, May 2016. European Language Resources Association (ELRA)
16. Adafre SF, de Rijke M (2006) Finding similar sentences across multiple languages in Wikipedia. In: Proceedings of the workshop on NEW TEXT Wikis and blogs and other dynamic text sources. https://www.aclweb.org/anthology/W06-2810
17. Patry A, Langlais P (2011) Identifying parallel documents from a large bilingual collection of texts: Application to parallel article extraction in Wikipedia. In: Proceedings of the 4th workshop on building and using comparable corpora: comparable corpora and the web, Portland, Oregon, June 2011, pp 87–95. http://www.aclweb.org/anthology/W11-1212
18. Tsou BK (2015) Augmented comparative corpora and monitoring corpus in Chinese: LIVAC and sketch search engine compared. In: Proceedings of the BUCC, Beijing, China, July 2015. Association for Computational Linguistics, pp 1–2. https://doi.org/10.18653/v1/W15-3401. https://www.aclweb.org/anthology/W15-3401
19. Munteanu DS, Marcu D (2006) Extracting parallel sub-sentential fragments from non-parallel corpora. In: Proceedings of the 21st international conference on computational linguistics and 44th annual meeting of the Association for Computational Linguistics, Sydney, Australia, July 2006. Association for Computational Linguistics, pp 81–88. https://doi.org/10.3115/1220175.1220186. https://www.aclweb.org/anthology/P06-1011
20. Rose T, Stevenson M, Whitehead M (2002) The reuters corpus volume 1-from yesterday's news to tomorrow's language resources. In: Proceedings of the third international conference on language

resources and evaluation (LREC), Las Palmas, Canary Islands—Spain, May 2002. European Language Resources Association (ELRA). http://www.lrec-conf.org/proceedings/lrec2002/pdf/80.pdf
21. Alexandre K, Ivan T, Binod B (2012) Inducing crosslingual distributed representations of words. In: Proceedings of the COLING, Mumbai, India
22. Hermann KM, Blunsom P (2014) Multilingual models for compositional distributed semantics. In: Proceedings of the 52nd annual meeting of the Association for Computational Linguistics, Baltimore, Maryland, June 2014. Association for Computational Linguistics, pp 58–68. https://doi.org/10.3115/v1/P14-1006. https://aclanthology.org/P14-1006
23. Schwenk H, Li X (2018) A corpus for multilingual document classification in eight languages. In: Proceedings of the eleventh international conference on language resources and evaluation (LREC 2018), Miyazaki, Japan, May 2018. European Language Resources Association (ELRA). https://aclanthology.org/L18-1560
24. Steinberger R, Pouliquen B, Van der Goot E (2013) An introduction to the Europe media monitor family of applications. arXiv:1309.5290
25. Sun Y, Zhu S, Yifan F, Mi C (2021) Parallel sentences mining with transfer learning in an unsupervised setting. In: Proceedings of the 2021 conference of the North American chapter of the Association for Computational Linguistics: student research workshop, Online, June 2021. Association for Computational Linguistics, pp 136–142. https://doi.org/10.18653/v1/2021.naacl-srw.17. https://aclanthology.org/2021.naacl-srw.17
26. Turney PD, Pantel P, Vector space models of semantics (2010) From frequency to meaning. J Artif Intell Res 37:141–188
27. Upadhyay S, Faruqui M, Dyer C, Roth D (2016) Cross-lingual models of word embeddings: an empirical comparison. In: Proceedings of the 54th annual meeting of the Association for Computational Linguistics, Berlin, Germany, Aug 2016. Association for Computational Linguistics, pp 1661–1670. https://doi.org/10.18653/v1/P16-1157. https://aclanthology.org/P16-1157
28. Chen J, Nie J-Y (2000) Automatic construction of parallel English-Chinese corpus for cross-language information retrieval. In: Sixth applied natural language processing conference, Seattle, Washington, USA, Apr 2000. Association for Computational Linguistics, pp 21–28. https://doi.org/10.3115/974147.974151. https://aclanthology.org/A00-1004
29. Resnik P, Smith N (2003) The web as a parallel corpus. Comput Linguist 29(3):349–380
30. Smith JR, Saint-Amand H, Plamada, PK, Callison-Burch C, Lopez A (2013) Dirt cheap Web-Scale parallel text from the Common Crawl. In: Proceedings of the ACL, Sofia, Bulgaria, Aug 2013, pp 1374–1383
31. Søgaard A, Ruder S, Vulić I (2018) On the limitations of unsupervised bilingual dictionary induction. In: Proceedings of the 56th annual meeting of the Association for Computational Linguistics (Volume 1: Long Papers), Melbourne, Australia, July 2018. Association for Computational Linguistics, pp 778–788
32. El-Kishky A, Chaudhary V, Guzman F, Koehn P (2019) A massive collection of cross-lingual web-document pairs. arXiv:1911.06154
33. Esplà M, Forcada M, Ramírez-Sánchez G, Hoang H (2019) ParaCrawl: web-scale parallel corpora for the languages of the EU. In: Proceedings of machine translation summit xvii volume 2: translator, project and user tracks, Dublin, Ireland, August 2019. European Association for Machine Translation, pp 118–119. https://www.aclweb.org/anthology/W19-6721
34. Etchegoyhen T, Azpeitia A (2016) Set-theoretic alignment for comparable corpora. In: Proceedings of the 54th annual meeting of the Association for Computational Linguistics (Volume 1: Long Papers), Berlin, Germany, Aug 2016. Association for Computational Linguistics, pp 2009–2018. https://doi.org/10.18653/v1/P16-1189. https://aclanthology.org/P16-1189

35. Ethayarajh K (2019) How contextual are contextualized word representations? Comparing the geometry of BERT, ELMo, and GPT-2 embeddings. In: Proceedings of the 2019 conference on empirical methods in natural language processing and the 9th international joint conference on natural language processing (EMNLP-IJCNLP), Hong Kong, China, November 2019. Association for Computational Linguistics, pp 55–65. https://doi.org/10.18653/v1/D19-1006. https://aclanthology.org/D19-1006
36. Fain M, Twomey N, Bollegala D (2021) Backretrieval: an image-pivoted evaluation metric for cross-lingual text representations without parallel corpora. arXiv:2105.04971
37. Cormode G, Muthukrishnan S, Sahinalp SC (2001) Permutation editing and matching via embeddings. In: International colloquium on automata, languages, and programming. Springer, pp 481–492
38. Cram D, Daille B (2016) Terminology extraction with term variant detection. In: Proceedings of ACL-2016 system demonstrations, Berlin, Germany, Aug 2016. Association for Computational Linguistics, pp 13–18. https://doi.org/10.18653/v1/P16-4003. https://aclanthology.org/P16-4003
39. Uszkoreit J, Ponte JM, Popat AC, Dubiner M (2010) Large scale parallel document mining for machine translation. In: Proceedings of the COLING '10: proceedings of the 23rd international conference on computational linguistics, pp 1101–1109
40. Utiyama M, Isahara H (2003) Reliable measures for aligning Japanese-English news articles and sentences. In: Proceedings of the 41st annual meeting of the Association for Computational Linguistics, Sapporo, Japan, July 2003. Association for Computational Linguistics, pp 72–79. https://doi.org/10.3115/1075096.1075106. https://aclanthology.org/P03-1010
41. Van der Maaten L, Hinton G (2008) Visualizing data using t-SNE. J Mach Learn Res 9:2579–2605
42. Vu T, Moschitti A (2021) CDA: a cost efficient content-based multilingual web document aligner. In: Proceedings of the 16th conference of the European chapter of the Association for Computational Linguistics: main volume, Online, Apr 2021. Association for Computational Linguistics, pp 3053–3061. https://aclanthology.org/2021.eacl-main.266
43. Jakubina L, Langlais P (2016) BAD LUC@WMT 2016: a bilingual document alignment platform based on Lucene. In: Proceedings of the first conference on machine translation: volume 2, Shared Task Papers, Berlin, Germany, Aug 2016. Association for Computational Linguistics, pp 703–709. https://doi.org/10.18653/v1/W16-2370. https://aclanthology.org/W16-2370
44. Laville M, Hazem A, Morin E, Langlais P (2020) Data selection for bilingual lexicon induction from specialized comparable corpora. In: Proceedings of the 28th international conference on computational linguistics, Barcelona, Spain (Online), Dec 2020. International Committee on Computational Linguistics, pp 6002–6012. https://www.aclweb.org/anthology/2020.coling-main.527
45. Levensthein VI (1966) Binary codes capable of correcting deletions, insertions, and reversals. Soviet Phys Doklady 10(8):707–710. https://nymity.ch/sybilhunting/pdf/Levenshtein1966a.pdf
46. Antonova A, Misyurev A (2011) Building a web-based parallel corpus and filtering out machine-translated text. In: Proceedings of the 4th workshop on building and using comparable corpora: comparable corpora and the web, Portland, Oregon, 2011. Association for Computational Linguistics, pp 136–144. https://aclanthology.org/W11-1218
47. Bhardwaj S, Hermelo DA, Langlais P, Bernier-Colborne G, Goutte C, Simard M (2020) Human or neural translation? In: Proceedings of the 28th international conference on computational linguistics, Barcelona, Spain (Online), Dec 2020. International Committee on Computational Linguistics, pp 6553–6564. https://doi.org/10.18653/v1/2020.coling-main.576. https://aclanthology.org/2020.coling-main.576
48. Biesialska M (2021) Major breakthroughs in unsupervised neural machine translation (v). https://mt.cs.upc.edu/2021/03/08/major-breakthroughs-in-unsupervised-neural-machine-translation-v/

49. Biś D, Podkorytov M, Liu X (2021) Too much in common: shifting of embeddings in transformer language models and its implications. In: Proceedings of the 2021 conference of the North American chapter of the Association for Computational Linguistics: human language technologies, Online, June 2021. Association for Computational Linguistics, pp 5117–5130. https://doi.org/10.18653/v1/2021.naacl-main.403. https://aclanthology.org/2021.naacl-main.403
50. Nguyen-Son H-Q, Thao T, Hidano S, Gupta I, Kiyomoto S (2021) Machine translated text detection through text similarity with round-trip translation. In: Proceedings of the 2021 conference of the North American chapter of the Association for Computational Linguistics: human language technologies, Online, June 2021. Association for Computational Linguistics, pp 5792–5797. https://doi.org/10.18653/v1/2021.naacl-main.462. https://aclanthology.org/2021.naacl-main.462
51. Nissim M, van Noord R, van der Goot R (2020) Fair is better than sensational: man is to doctor as woman is to doctor. Computat Linguist 46(2):487–497
52. Mikolov T, Le QV, Sutskever I (2013) Exploiting similarities among languages for machine translation. arXiv:1309.4168
53. Lample G, Conneau A, Ranzato MA, Denoyer L, Jégou H (2018) Word translation without parallel data. In: Proceedings of the international conference on learning representations
54. Laroche A, Langlais P (2010) Revisiting context-based projection methods for term-translation spotting in comparable corpora. In: Proceedings of the 23rd international conference on computational linguistics (Coling 2010), pp 617–625, Beijing, China, Aug 2010. Coling 2010 Organizing Committee. https://aclanthology.org/C10-1070
55. Sharoff S (2006) Open-source corpora: using the net to fish for linguistic data. Int J Corpus Linguist 11(4):435–462
56. Baroni M, Bernardini S, Ferraresi A, Zanchetta E (2009) The WaCky wide web: a collection of very large linguistically processed web-crawled corpora. Lang Res Eval 43(3):209–226
57. Baroni M, Dinu G, Kruszewski G (2014) Don't count, predict! A systematic comparison of context-counting versus context-predicting semantic vectors. In: Proceedings of the 52nd annual meeting of the Association for Computational Linguistics, Baltimore, MD, June 2014
58. Bateman JA (1997) Enabling technology for multilingual natural language generation: the KPML development environment. Natural Lang Eng 3(1):15–55
59. Bengio Y (2012) Deep learning of representations for unsupervised and transfer learning. In: Guyon I, Dror G, Lemaire V, Taylor G, Silver D (eds), Proceedings of ICML workshop on unsupervised and transfer learning, volume 27 of Proceedings of machine learning research, Bellevue, Washington, USA, 02 Jul 2012. PMLR, pp 17–36. https://proceedings.mlr.press/v27/bengio12a.html
60. Bernardini S, Ferraresi A (2013) Old needs, new solutions: comparable corpora for language professionals. In: Sharoff S, Rapp R, Zweigenbaum P, Fung P (eds), Building and using comparable corpora. Springer, pp 303–319
61. Schäfer R, Bildhauer F (2013) Web corpus construction. Synthesis lectures on human language technologies. Morgan & Claypool Publishers
62. Wenzek G, Lachaux M-A, Conneau A, Chaudhary V, Guzman F, Joulin A, Grave E (2019) CCNET: extracting high quality monolingual datasets from Web crawl data. arXiv:1911.00359
63. Wu D (1994) Aligning a parallel English-Chinese corpus statistically with lexical criteria. In: Proceedings of the 32nd annual meeting on Association for Computational Linguistics (ACL '94), Stroudsburg, PA, USA. Association for Computational Linguistics, pp 80–87
64. Wu S, Dredze M (2019) Beto, Bentz, Becas: the surprising cross-lingual effectiveness of BERT. arXiv:1904.09077
65. Wu S, Dredze M (2020) Are all languages created equal in multilingual BERT? In: Proceedings of the 5th workshop on representation learning for NLP, Online, July 2020. Association

for Computational Linguistics, pp 120–130. https://doi.org/10.18653/v1/2020.repl4nlp-1.16. https://aclanthology.org/2020.repl4nlp-1.16
66. Xiao R (ed) (2020) Using corpora in contrastive and translation studies. Cambridge Scholars Publishing
67. Xu H, Koehn P (2021) Cross-lingual BERT contextual embedding space mapping with isotropic and isometric conditions. arXiv:2107.09186
68. Conneau A, Khandelwal K, Goyal N, Chaudhary V, Wenzek G, Guzmán F, Grave E, Ott M, Zettlemoyer L, Stoyanov V (2019) Unsupervised cross-lingual representation learning at scale. arXiv:1911.02116
69. Sharoff S (2020) Know thy corpus! robust methods for digital curation of web corpora. In: Proceedings of the 12th language resources and evaluation conference, pp 2453–2460, Marseille, France, May 2020a. European Language Resources Association. ISBN 979-10-95546-34-4. https://aclanthology.org/2020.lrec-1.298
70. Blei DM, Ng AY, Jordan MI (2003) Latent Dirichlet allocation. J Mach Learn Res 3:993–1022
71. Dieng AB, Ruiz FJR, Blei DM (2020) Topic modeling in embedding spaces. Trans Assoc Comput Linguist 8:439–453. https://doi.org/10.1162/tacla00325. https://www.aclweb.org/anthology/2020.tacl-1.29
72. Fothergill R, Cook P, Baldwin T (2016) Evaluating a topic modelling approach to measuring corpus similarity. In: Proceedings of the tenth international conference on language resources and evaluation (LREC'16), Portorož, Slovenia, May 2016. European Language Resources Association (ELRA), pp 273–279
73. Radford A, Wu J, Child R, Luan D, Amodei D, Sutskever I (2019) Language models are unsupervised multitask learners. OpenAI Blog 1(8)
74. Kunilovskaya M, Sharoff S (2019) Towards functionally similar corpus resources for translation. In: Proceedings of the international conference on recent advances in natural language processing (RANLP 2019), Varna, Bulgaria, Sept 2019. INCOMA Ltd, pp 583–592. https://doi.org/10.26615/978-954-452-056-4_9. https://aclanthology.org/R19-1069
75. Schwenk H, Chaudhary V, Sun S, Gong H, Guzmán F (2021) WikiMatrix: mining 135M parallel sentences in 1620 language pairs from Wikipedia. In: Proceedings of the 16th conference of the European chapter of the Association for Computational Linguistics: main volume, Online, Apr 2021. Association for Computational Linguistics, pp 1351–1361. https://doi.org/10.18653/v1/2021.eacl-main.115. https://aclanthology.org/2021.eacl-main.115

Extraction of Parallel Sentences

4

As explained in Chap. 1 and later developed in Chap. 6, Machine Translation (MT) engines need to be trained with large numbers of parallel sentences or segments. The quantity and diversity of existing parallel text is limited however. This motivates the search for parallel sentences in comparable corpora. By exploring a larger share of the levels of comparability introduced in Sect. 1.2, a much larger source of multilingual data can be obtained. Strongly comparable corpora such as Wikipedia entries [1, 62] or news text [2] are rife with parallel sentences and have been among the first to be explored. Even weakly comparable corpora such as texts selected according to similar criteria across multiple languages are bound to contain parallel sentences or segments. Finding good-quality parallel sentences is likely to be more difficult though as one moves towards gradually less comparable text. The extracted parallel sentences have been shown to improve machine translation quality in a number of cases [3, 64]. This is exemplified further in Chaps. 6 and 7.

In this chapter, when referring to two sentences in two languages, we use a number of terms interchangeably such as *parallel sentences*, *translation equivalents*, *cross-lingual equivalents*, to refer to sentence pairs whose meanings are sufficiently similar to be used as translations of each other. Note that in machine translation the basic *translation unit* is generally a sentence, but can depart from it to some extent because of the variety of text structure and the resulting complexity of sentence segmentation. Before focusing on non-parallel, comparable corpora, we begin with a summary of parallel sentence extraction from parallel corpora (Sect. 4.1). We then examine how translation equivalence is usually measured (Sect. 4.2) and present datasets to test and evaluate parallel sentence extraction systems (Sect. 4.3). We introduce the general principles of parallel sentence extraction in comparable corpora (Sect. 4.4), and detail a number of more specific methods, distributing them into non-neural methods (Sect. 4.5), supervised neural methods (Sect. 4.6), and unsupervised neural methods (Sect. 4.8).

S. Sharoff et al., *Building and Using Comparable Corpora for Multilingual Natural Language Processing*, Synthesis Lectures on Human Language Technologies,
https://doi.org/10.1007/978-3-031-31384-4_4

4.1 Extraction from Parallel Corpora

The detection of parallel sentences in a bilingual corpus was first addressed in the context of parallel corpora made of texts in a source language and their translations in a target language. The task in that context was to identify sentences that are translations of each other, by exploring the sequence of sentences in the source text and the sequence of sentences in the target text. That situation creates a task of sequence alignment, hence the term 'sentence alignment' [4]. A key property of translated text is that the order of sentences in a translated text is usually kept unchanged. In parallel texts, sentence alignment must therefore find which source sentences align to which target sentences in the order in which they are written in the texts. Most algorithms that align sentences in parallel corpora can thus rely on that monotonous order and look for additional, usually weak clues that may confirm that two sentences are similar. These clues include sentence length [5] and sentence contents based upon words or characters [6], as well as a number of additional heuristics [7]. Bootstrapping methods have been designed that rely on length-based alignment to extract high-confidence parallel sentences. These parallel sentences create a parallel subcorpus from which a basic bilingual dictionary [8] or machine translation system [9] can be obtained. The obtained resource is then used as additional information to align the remaining sentences. Finally, [10] designed a parallel sentence alignment method that relies on cross-language information retrieval techniques, thus does not require translated text to preserve sentence order. This method can be seen as laying out the principles used later in comparable corpora and detailed from Sect. 4.4 on.

The use of sentence length [5] assumes that long source language sentences are typically translated by long target language sentences, and short source language sentences by short target language sentences. Given a source language corpus and its target language translation, among many potential alignments the one is considered optimal which minimizes the length differences between aligned sentences. Figures 4.1 and 4.2 illustrate this for a toy corpus of three German sentences whose translation comprises four English sentences. Sentence length is computed as the number of characters in a sentence, and the alignment where the sum of the absolute length differences between aligned sentences is minimal is considered to be the best one. In our example, this is the case for the first alignment. In practice, given good quality translations with no major re-arrangements concerning sentence order, this method can achieve accuracies of above 99% correctly aligned sentences.

As an alternative approach, if a basic bilingual dictionary is available, instead of looking at sentence lengths it is also possible to look at the number of matches of dictionary entries between aligned sentences, as illustrated in Fig. 4.3. Among the three alignments shown, alignment 1 would be considered best as, with seven matches, it has the highest number of dictionary hits.

A third way to evaluate sentence alignments is by looking at substrings as shown in Fig. 4.4. This works best for related languages where words are often spelled similarly

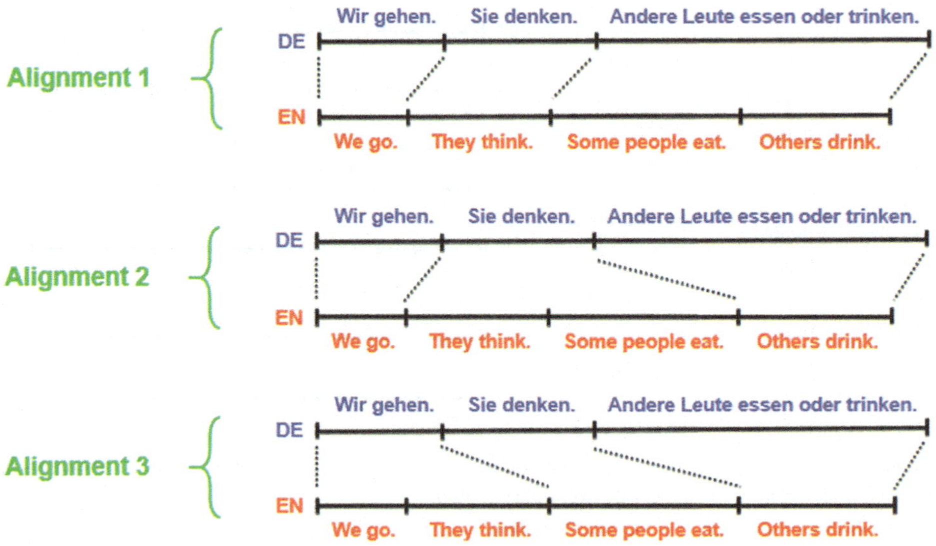

Fig. 4.1 Three possible alignments of a German–English toy corpus

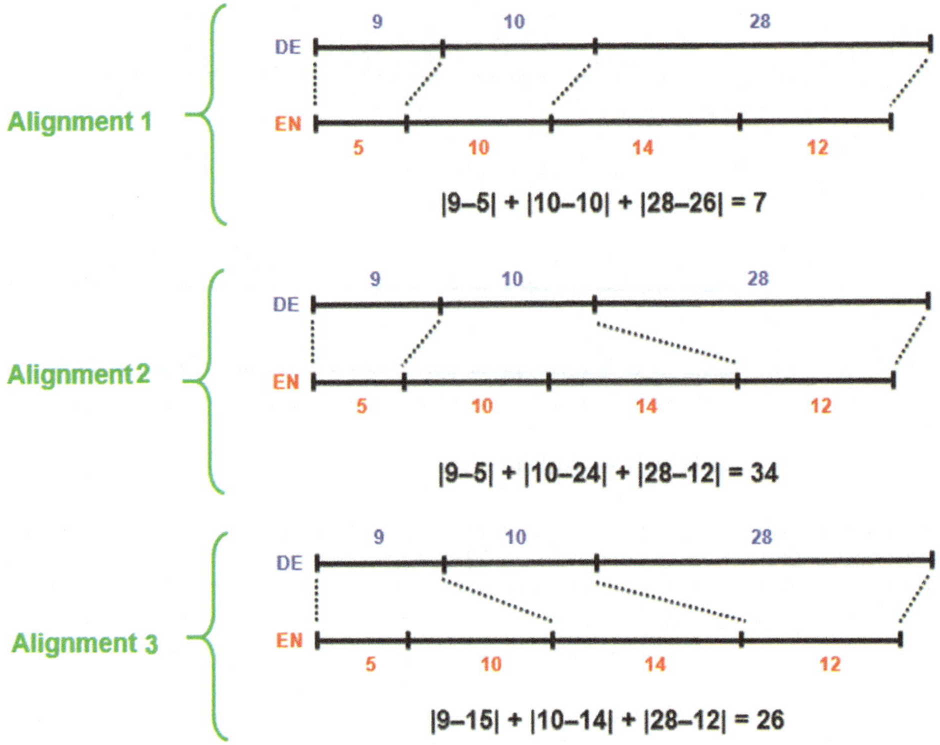

Fig. 4.2 Evaluation of three possible alignments

German	English
andere	other
denken	think
essen	eat
gehen	go
Leute	people
sie	they
oder	or
trinken	drink
wir	we

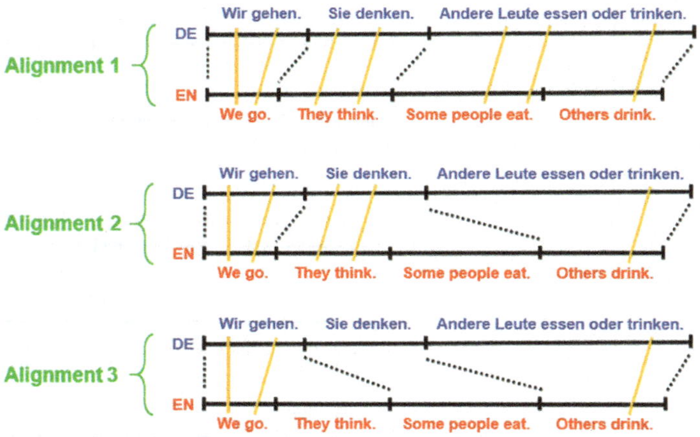

Fig. 4.3 Sentence alignment using a basic dictionary (of which the relevant extract is shown on the left side). The yellow lines indicate dictionary matches

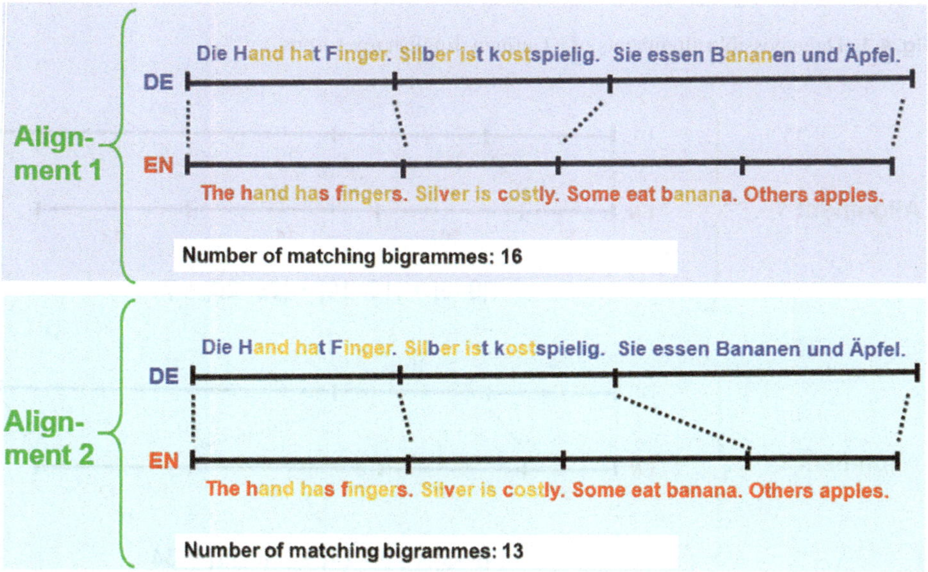

Fig. 4.4 Sentence alignment based on the number of matching character pairs (bigrams). Substrings which are identical in the source and in the target language are shown in yellow. To give an example, our counting scheme identifies three matching bigrams between "Silber" and "Silver" as "Si", "il" and "er" occur in both strings

(cognates). For unrelated languages such as English–Chinese it may still work to some extent if there are enough proper names or foreign citations in the translation.

The remainder of this chapter examines parallel sentence extraction in non-parallel, comparable corpora.

4.2 Assessing Cross-Lingual Sentence Similarity

The previous section examined parallel texts, in which sentence alignment can usually rely on the parallel ordering exhibited by sentences in translated texts and use weak signals to identify an optimal sentence alignment.

There is no such parallel order in comparable corpora. Instead, sentences are deemed parallel based on their intrinsic properties. Since the main purpose of spotting parallel sentences is to train machine translation systems (see Chap. 6), two sentences are deemed parallel if they can be considered good translations of each other. The usual criteria for judging the quality of translations include adequacy and fluency. Adequacy measures whether the information in the source sentence is preserved in the target sentence, whereas fluency measures whether the target sentence is well formed and fluent. In the evaluation of parallel sentence extraction methods, there is a tacit assumption that the comparable corpora involved in the evaluation are made of well-formed sentences, which leads to a focus on adequacy only[1]. Hence the key criterion used to assess the quality of a sentence pair found in comparable corpora: do the two sentences carry the same meaning?

When asking humans to evaluate sentence pairs to prepare a dataset or to judge a system's output, instructions must be given that enable the evaluators to contrast positive cases (the two sentences carry the same meaning) and negative cases (the two sentences carry different meanings). Guidelines for the evaluation of cross-lingual semantic text similarity [12] (STS 2016) can help to examine this question. Human evaluators were given a sentence pair and asked to give a rating on a Likert scale from 0 (worst) to 5 (best). Level 5 (*The two sentences are completely equivalent, as they mean the same thing*) corresponds to an exact or close-to-exact translation, level 4 (*The two sentences are mostly equivalent, but some unimportant details differ*) to a less good but maybe acceptable translation, whereas level 3 (*The two sentences are roughly equivalent, but some important information differs or is missing*) would be an inexact translation and would probably be much less suitable to train a machine translation system. The remaining levels are clearly unsuitable: level 2 (*The two sentences are not equivalent, but share some details*), level 1 (*The two sentences are not equivalent, but are on the same topic*) and level 0 (*The two sentences are on different topics*).

Given the above-mentioned STS 2016 scale, guidelines for parallel sentence extraction from comparable corpora can be defined according to the requirements of the downstream

[1] See however [11], who performed an audit of some web-mined text collections and uncovered a number of quality issues.

task by setting a threshold on that scale. If a high semantic similarity is required, level 5 will be set as the threshold; if a less high semantic similarity is allowed, level 4 (or even 3) may be selected instead.

4.3 Datasets and Evaluation

We present here datasets that can be of use when evaluating systems that aim to detect parallel sentence pairs in comparable corpora. As explained in the previous section, in the end the quality of a parallel sentence pair selected by a system is evaluated by the level of semantic similarity of the two sentences.

However, a system that aims to extract parallel sentence pairs from a pair of comparable corpora must be prepared to find matching pairs among large numbers of candidate sentences in each of the two corpora. The large number of candidate target sentences makes the selection of correct pairs much more challenging, since more false friends may be present. Moreover, it also creates computational challenges because of the quadratic complexity of the search for sentence pairs in the two corpora. Datasets with larger numbers of both source and target candidate sentences are therefore better approximations of an ecological situation.

Another dimension is the variety of languages that a system can handle. These may include better-resourced languages or under-resourced languages, languages with various alphabets, syllabaries, or logographic writing systems, with different tokenization or morphology (see also Sect. 3.2). These differences are possible sources of difficulty when processing these languages. The number and typological variety of the languages in a dataset are therefore important features of a dataset.

The following datasets include graded or binary semantic similarity judgments; their numbers of source and target sentences differ, as well of the numbers of languages they contain:

SemEval 2016 Cross-lingual Semantic Text Similarity (STS 2016): Agirre et al. [12] Source and target sentences are already paired, only the semantic similarity capability of systems is assessed; about 3,300 pairs are in the test set, among which 600 were judged; one pair of languages is included (English-Spanish); the gold standard contains judgments on a 6-point Likert scale, evaluation is by Pearson correlation with gold standard labels. This dataset is really suited for cross-language similarity evaluation.

BUCC 2017–2018 Parallel Sentence Extraction (BUCC 2017–2018): Zweigenbaum et al. [13] Independent source sentences and target sentences are provided, in the order of 10^5 for both source and target for each of training and test; parallel pairs amount to about 2% of these numbers; 4 pairs of languages are included (English paired to German, French, Russian, and Chinese), which represent three different writing systems and four language families; the gold standard contains binary judgments, evaluation is by F1 score of system sentence pairs. This dataset was specifically designed to evaluate parallel sentence extraction systems and is of interest to test the scalability of parallel sentence extraction systems. The same shared task was run in both 2017 and 2018, hence the equivalent names BUCC 2017 or BUCC 2018 for this corpus.

Tatoeba Multilingual Similarity Search: Artetxe and Schwenk [14] Source and target sentences are provided separately, in the order of 10^3 for 72 languages and down to 10^2 for the rest; each source or target sentence has a parallel counterpart; 93 pairs of languages are provided (English paired to 93 different languages), in more than 20 different families and over 12 different writing systems; the gold standard contains binary judgments; given a source sentence, evaluation is by error rate of the top-1 returned target sentence. This dataset is of particular interest to test the language-wise scalability of parallel sentence extraction systems.

The XTREME benchmark [15], a composite benchmark for multi-lingual learning, includes both the BUCC 2018 training data and the Tatoeba dataset.

4.4 General Principles

Parallel sentences can be spotted in comparable corpora by using external and internal cues. The former are typically used in a pre-filtering step to focus the application of the latter on more relevant material.

External methods examine the texts in which two sentences occur to control their comparability: are these texts similar [24, 25]? For instance, do they cover the same specific events or subject domain, were they authored in the same time period [2], do they belong to the same text genre? This type of test focuses the search on corpora and texts that are more likely to contain parallel sentences.

Internal methods examine the contents of a sentence pair to determine whether the two sentences are equivalent cross-linguistically. For instance, do they contain word pairs that are translations of each other [2, 26], are their lengths similar [2, 26], are their vector representations close to one another [27, 28]? This notion of cross-language sentence similarity is the core criterion that must be examined, and all methods ultimately rely on it at some point.

Given two comparable corpora that each contain a large number of sentences, exhaustively assessing the similarity of the full cross-product of their sentence pairs is a quadratic problem that may be computationally prohibitive. Many methods therefore decompose the task into two phases that use two kinds of methods. The distinction between these two types of methods is indeed relative.

In the first phase, scalable filtering methods are applied to select a subset of sentence pairs that are more promising candidates for being parallel. Cross-language information retrieval methods [10] are often used in this phase. The cross-language, i.e., translation step, can be restricted to simple dictionary-based word translation [2, 26] or use full-fledged machine translation of sentences [32, 63]. The information retrieval methods include BM25 [24], tf.idf [2, 63], possibly within search engines such as Lucene [33, 66].

The second phase can then apply more computationally demanding methods to the resulting subset.

4.5 Pre-neural Methods

In these non-neural methods, words are the basic representation unit, possibly used in sparse vectors, and word or sentence translation models are required to bridge the cross-language gap. These methods also spend a large share of their effort on external clues.

Early work on the extraction of parallel sentences from comparable corpora includes [2, 24–26, 65]. Munteanu and Marcu [2] have set the model for much of the later work. They start from large monolingual corpora composed of news articles in Arabic, Chinese, and English and from a probabilistic bilingual dictionary for each studied language pair. In a first filtering step based upon external clues, to increase the chances of finding parallel sentences, they select similar news articles in the same time frame across languages. More precisely, for each document in language L_2 (Arabic or Chinese), they build a pseudo-English document made of the top 5 translations of its words through the dictionary. This document is used to retrieve, through an information retrieval toolkit, the top 20 English documents among those published within ± 5 days of the L_2 document's date.

Within the resulting matching document pairs, an additional filtering step based upon internal clues is then applied to sentence pairs. First, the ratio of sentence lengths must not be greater than two. Second, at least half of the words in a sentence must have a translation in the other sentence.

The final selection of parallel sentence pairs is obtained by training a maximum-entropy classifier with the following features. Sentence length features include lengths, difference and ratio; word translation features include the percentage of words in a sentence that have a translation in the other sentence through the dictionary; word alignment features include the percentage and number of words that have no connection, top 3 largest 'fertilities' (n in 1:n alignments), length of longest continuous span of aligned words, and length of longest un-connected substring. Word alignment is obtained through five versions of a modified IBM-1 model [34] trained on parallel corpora, hence producing five sets of word alignment features.

Often found in both these early explorations and later work are:

- the notion of document similarity based upon information retrieval: BM25 [24], tf.idf [2, 63], cosine similarity [25], often based upon information retrieval engines such as Lucene [33, 66];

- the use of cross-language word overlap in sentence pairs [2, 3, 24, 33]. This comes in many flavors; for instance, given a candidate sentence pair in which the word translation model has identified word translations in the other sentence, [33] consider as additional translations the following three types of words: word pairs that share a common initial substring longer than 3 characters, capitalized words, and numbers;
- sentence length features [2, 3, 26];
- statistical translation models including word-level IBM-1 [2, 3, 26, 33, 66], IBM-4 [25], or fuller translation models [32, 63];
- the final sentence-alignment decision, using scores or supervised binary classifiers such as maximum entropy [2, 3]. In matched comparable documents such as Wikipedia pages, where a global sentence alignment model may be relevant to identify translated sentence clusters, [3] use a Conditional Random Fields sequence classifier to take sentence context into account.

Similar principles have been included in work on the more specific task of parallel sentence cleaning, e.g. in [67].

The methods in this section rely on a number of heuristics and features whose selection and parameters were carefully chosen by their authors, and some of the publications report ablation experiments that show their contribution. To obtain optimal results for new datasets though, these choices would ideally need to be tuned on a development set through some form of grid search.

4.6 Supervised Neural Methods

Supervised neural methods for parallel sentence extraction use bilingual supervision signals that rely on cross-lingual vector spaces (see Sect. 2.1), also called cross-lingual embeddings. Most of the time, these signals are obtained from an initial set of parallel sentences that are used to derive either multilingual word embeddings [37] or multilingual sentence embeddings [38].

Some neural parallel sentence extraction methods replace the features and classifiers of the above-described pre-neural methods (Sect. 4.5) with multilingual embeddings and neural classifiers. This is the case of [37], who first learn bilingual word embeddings from parallel sentences then use them as input features to a convolutional neural network to classify sentence pairs as parallel or not.

Another set of methods, such as those implemented in the LASER toolkit[2] [14, 27, 38, 39], compute embeddings for the input sentences then rely on the similarity of these embeddings to decide whether or not two input sentences are parallel. The more similar the embeddings, the higher the similarity of the two sentences, hence the more likely they are to be translations of each other. Then a similarity threshold can be set above which a sentence

[2] https://github.com/facebookresearch/LASER.

pair is considered to be parallel. For instance, [27] uses cosine similarity and optimizes this threshold on the BUCC 2018 training sets: thresholds from 0.50 to 0.64 are selected. In contrast, given a source sentence x, [39] take as a score the *margin* between the cosine of a given candidate target sentence y and the average cosine obtained with the k nearest neighbors NN() of x and of y:

$$\text{score}(x, y) = \text{margin}\left(cos(x, y), \frac{1}{2}\left(\sum_{z\in \text{NN}_k(x)} \frac{cos(x, z)}{k} + \sum_{z\in \text{NN}_k(y)} \frac{cos(y, z)}{k}\right)\right) \tag{4.1}$$

They find that the best margin function for the BUCC 2018 dataset is the *ratio* $margin(a, b) = \frac{a}{b}$ and $k = 4$: it increases the F1-score by 0.3pt on the training set compared to the *distance* $margin(a, b) = a - b$, and both gain more than 10pt F1-score over using the raw cosine similarity. A threshold is set for that margin score: it is set to 1.02 in [62] and to 1.06 for the much larger dataset in [64].

At the intersection of these two families of neural parallel sentence extraction methods, [40] use a siamese architecture to train both two BiRNN sentence encoders and a sentence pair classifier. The dual encoder model of the LABSE sentence embeddings [71] follows this path, using pre-trained BERT-based sentence encoders.

To compute monolingual then multilingual sentence embeddings, the following neural methods have been proposed:

- learning individual word embeddings in a self-supervised way, then averaging them, possibly with tf.idf weighting as in [16];
- learning sentence embeddings directly through recurrent neural networks (RNN) or Transformer networks, also in a self-supervised way. For instance, [41] trained an RNN-based sequence-to-sequence encoder-decoder model to reconstruct the previous and next sentences given the current sentence, then used the output of the encoder to represent new sentences.

The key point for cross-language sentence similarity is to design a representation space that is shared across the compared languages.

From monolingual to multilingual embeddings. Creating a shared representation space can be performed by learning a mapping from each language to that shared representation space (which can be that of one of the languages), using word pairs in a bilingual dictionary as training examples. As in many other contexts, the bilingual dictionary can also be obtained from a parallel corpus. Section 5.5 provides more detail on the alignment of word embeddings.

Direct creation of multilingual sentence embeddings. An alternative method consists in learning the shared representation space directly from a corpus of parallel sentences, using the encoder output of a multilingual, encoder-decoder neural machine translation system.

The internal representations produced by the encoder of a neural machine translation system have been shown to be beneficial for monolingual natural language processing tasks [47]. Furthermore, because the same system is trained on multiple source or target languages, the internal representations produced by the encoder tend to be interlingual and make sentences with similar meanings, including translations, closer in the representation space. Schwenk and Douze [38] explored BiLSTM architectures with multiple encoders and decoders. Sentence embeddings are obtained by max-pooling of the encoder outputs. The decoders are only used at training time. Their one-to-many strategy, given an N-way parallel corpus, translates from one source language simultaneously to the set of N-1 other languages. Nevertheless, since the source and target languages are changed at each mini-batch, every source language shares multiple target languages, and every target language shares multiple source languages. Their many-to-one strategy uses multiple languages as input at the same time and one language as output, and the multiple input sentence embeddings are averaged to feed the decoder. They observed, among other conclusions, that one-to-many language translation architectures performed better than many-to-one architectures. However, because the one-to-many strategy in [38] requires an N-way parallel corpus, [14] later simplified it to a one-to-two strategy with English and Spanish as target languages. They simultaneously trained a BiLSTM encoder-decoder architecture on 93 source languages and these two target languages, so that the output of this single encoder can be used to represent a sentence in any of the source languages. España-Bonet et al. [70] studied the evolution of GRU-based encoder representations during training, using the sum of encoder outputs to compute sentence representations. They concluded that the best interlingual representation for spotting translations was obtained much before the optimal training point for machine translation. Section 6.2 provides additional information on sentence embedding methods.

We noted earlier that typical parallel sentence extraction methods begin by reducing the (initially quadratic) size of the set of candidate sentence pairs that they will examine precisely, using scalable methods such as those of Information Retrieval. Some neural methods for parallel sentence extraction nevertheless dispense from such a filtering phase. The quadratic step in these methods is the search for the most similar target sentence embeddings for source sentences. This is an instance of nearest neighbor search, for which efficient algorithms exist [22] and have been implemented, e.g., in the Faiss library for efficient similarity search and clustering of dense vectors [23]. This makes it feasible for [27] to search for the k nearest sentence representations from 144 million English sentences to 187 million German sentences, a total of $2.7\ 10^{16}$ distance calculations.

The methods in this section were applied at scale to create very large parallel sentence corpora in many different languages. To create the WikiMatrix dataset, [62] use the LASER multilingual sentence embeddings computed by [14] for 93 languages to mine parallel sentences in Wikipedia. For 1620 language pairs covering 96 languages, they were able to extract at least 10k sentences, for a total of 135 million parallel sentences. These parallel sentences include over 100 million with two languages other than English, hence their interest to train machine translation systems for distant languages without pivoting through English.

In a similar way, [64] created the CCMatrix dataset by collecting 10.8 billion parallel sentences in 90 languages from CommonCrawl snapshots, among which less than 3 billion involve English. In the mining process, they organized all languages hierarchically into 12 groups based upon language families and geographic groupings, and selected 27 *bridge languages* as major languages in each group. For the sake of efficiency, instead of the full cross-product of 4005 language pairs, mining only considered the 716 language pairs within one group or within the set of bridge languages. For many languages, CCMatrix provides sentence pairs with multiple target languages: this departs from the usual prevalence of English-aligned bitexts. This facilitates the development of multilingual machine translation systems: [68] used the CCMatrix dataset to train a massively multilingual machine translation system that can translate directly between any pair among 100 languages. This system, trained from parallel sentences collected from CommonCrawl as explained above, improves BLEU scores in 13 out of 18 comparisons on English-centric WMT datasets against the best published single models, with a average of +1.9pt. It also improves average BLEU scores by 1.3 to 4.3pt over three other multilingual models.

In a more focused experiment, [71] use the BERT-based LABSE multilingual sentence embeddings to extract parallel sentences from CommonCrawl in the English-to-Chinese and English-to-German language pairs and train two bilingual machine translation systems.

4.7 Limitations of Supervised Methods in Low-Resource Settings

The supervised methods presented in Sect. 4.6 require parallel training data. Most of these methods are trained on large parallel corpora: 500k parallel sentence pairs in [40], 8.3M six-way parallel sentences in [38], up to 9.3M parallel sentence pairs per language in [14], up to 200M parallel sentence pairs per language in [71]. This is a drawback in low-resource scenarios, for which comparable corpora are presented as a solution.

Artetxe and Schwenk [14] address a very large number of languages, which include some with few or no training data: this study illustrates how parallel sentence extraction methods fare in low-resource scenarios. The authors train LASER sentence embeddings on 93 languages with a number of parallel sentence pairs ranging from 500 to 2k (Coastal Kadazan, Khmer, Irish, Chabacano, Kadazan Dusun, Meadow Mari, Cornish, Lingua Franca Nova, Wu) to 8–9M (Danish, Arabic, Italian, German, French, Russian).

One of the evaluations they perform tests the identification of parallel sentences obtained from the Tatoeba dataset (see Sect. 4.3) between English and one of 112 languages, using nearest neighbor search based upon the above-mentioned LASER sentence embeddings. Depending on the language and the size of its training data, sentence similarity error rates vary from a very low 1.10 (Norwegian Bokmål) to a very high 95.80 (Irish).

Our own analysis of the results in [14] confirms some of the author's comments. On the one hand, languages with a small number of training examples tend to incur a high error rate. On the other hand, languages related to other languages with many training examples benefit from the joint training of the encoder. The combination of these two tendencies can be seen in the results of the 29 languages with no training data at all (zero shot) and of the 9 languages with no more than 2000 training sentences (few shot).

24 out of 29 zero-shot languages have error rates above 42%, whereas the other 5 have error rates below 34%. These 5 languages are all related to well-represented languages: Asturian is a West Iberian Romance language, Norwegian Nynorsk is one of the two written standards of the Norwegian language, Faroese is a North Germanic language, Egyptian Arabic is the most widely spoken vernacular Arabic dialect in Egypt, and Novial is a constructed international auxiliary language whose vocabulary is based largely on the Germanic and Romance languages while its grammar is influenced by English.

Among the 9 few-shot languages, 3 have an error rate below 36%, and are all related to resource-rich languages: these are Chabacano, a group of Spanish-based creole language varieties, Wu, a group of Sinitic languages, and Lingua Franca Nova, again an auxiliary constructed language, whose vocabulary is based primarily on the Romance languages. Conversely, the other 6 languages, which yield an error rate above 77%, do not have closely related, resource-rich languages in the training set to compensate for their small amount of training data. These are Irish (Scottish Gaelic and Manx are absent from the training data); Kadazan Dusun and Coastal Kadazan, two dialects spoken in the Sabah state of Malaysia in Borneo; Cornish, a Southwestern Brittonic language (its two contemporary relatives are Welsh, absent from the training data, and Breton which, although with 29k training examples, obtains a high error rate); Meadow Mari, a dialect of Mari, an Uralic language; and Khmer, an Austroasiatic language that is considered alone in its branch.

Designing multilingual training sets with language families in mind may therefore be a direction to better cope with low-resource situations. This observation recurs in other chapters of this book, and is emphasized in Sect. 7.1.2.

4.8 Unsupervised Neural Methods

The methods described in the previous sections rely on strong bilingual signals obtained either from existing bilingual dictionaries or from parallel corpora.

Dispensing from these bilingual signals remains a quest that researchers have been actively following, aiming to design unsupervised methods that create bilingual word or sentence embeddings from comparable corpora. These bilingual embedding spaces are then leveraged to identify word translations (see Sect. 5.5) and parallel sentences.

As explained in more detail later in Sect. 5.5, most of these methods align independently created monolingual embedding spaces. Artetxe et al. [48] initialize the embedding alignment process with often available signals: identical strings found in both corpora, such as numbers and punctuation marks; the method then iteratively expands the number of aligned word embeddings in the two monolingual spaces. Lample et al. [49] perform the initialization through adversarial training. Alvarez-Melis and Jaakkola [51] model the problem as optimal preservation of distances between pairs of points in the two monolingual spaces. In contrast, [52, 69] rely on the mBERT contextual multilingual embeddings directly obtained by pre-training the BERT model [53] on Wikipedia sentences across 104 languages and released with BERT.

To address parallel sentence extraction, [28] first rely on the MUSE system of [49] to create unsupervised bilingual word embeddings. They use these embeddings to create word translation dictionaries. The dictionary for a source word contains its top 100 nearest target-language words according to their word embedding similarity, as well as its most similar target-language words according to their normalized Levenshtein distance (a similarity is obtained as one minus the distance). The target words are weighted by their similarity.

Then, they test two methods to score a pair of candidate parallel sentences. First, they take the cosine similarity of sentence embeddings obtained by averaging the words of each sentence after removal of stopwords, digits and punctuation.

Second, they also greedily align source and target words based upon their similarity scores, and compute the sentence similarity score as the average similarity of aligned word pairs.

Pursuing this thread, [54] additionally use Cross-Domain Similarity Local Scaling (CSLS) [49] to compute word embedding similarity. To increase the precision of parallel sentence detection, they require the presence of continous segments of aligned words with similarity scores above a set threshold and apply additional heuristics. The presence of aligned segments and their similarity scores are finally used to compute a similarity score for a sentence pair.

These methods are compared on the English-German, English-French and English-Russian language pairs of the BUCC 2017 parallel sentence extraction dataset [13] (see Sect. 4.3 above). Word alignment in Hangya et al. [28] dramatically improves recall in all cases, albeit at the expense of precision for en-de. The Levenshtein distance improves both recall and precision for en-de and en-fr, but not for en-ru because of their different writing systems. The additional parallel segment requirement in Hangya and Fraser [54] boosts precision in all cases, but decreases recall, bringing a strong increase in F1 score in en-de and en-ru, although a small decrease in F1 score in en-fr.

Hangya and Fraser [54] also evaluate the contribution of parallel sentences mined with this method to the unsupervised neural machine translation system of Artetxe et al. [55] on the WMT 14 en-de and WMT 16 en-fr datasets. Rounds of training on the parallel sentence pairs obtained by the last method (with parallel segment detection) are inserted in the training procedure of that unsupervised NMT system: this brings an improvement of a few points in BLEU scores.

Moving to the contextual word embeddings of the BERT transformer models [53, 69] create sentence embeddings by processing them with the multilingual mBERT model and mean-pooling the output word representations. Given two monolingual corpora, they compare the embedding of each source sentence with that of all target sentences and keep its $k = 4$ nearest neighbors. They then calibrate the cosine similarity of two sentence embeddings x and y by dividing it by the average cosine similarities of x and y with each of their four nearest neighbors, according to the ratio margin score of [39] (see Eq. 4.1). They filter out sentence pairs whose calibrated similarity is below a set threshold, or which contain digit sequences that do not match, or are too similar at the character level (which might reveal citations from one language into the other language).

They fine-tune these representations by self-training mBERT on the obtained sentence pairs to detect parallel sentences: the nearest neighbor, if kept after filtering, is considered as a positive example, and the next $k - 1$ neighbors are used as negative examples. The updated mBERT model is then used to recompute the nearest neighbors and select updated parallel sentences. Several such iterations can be performed if relevant.

They evaluate these methods on all four language pairs of the BUCC 2017 dataset: English-German, English-French, English-Russian, and English-Chinese. The initial mBERT mean-pooled embeddings yield F1 scores close to that of [54] on en-de (−1pt) and en-fr (+2pt) and 12 points higher for en-ru. Adding the filtering steps further improves by 2–5pt, and self-training again leads to an increase of 8–14pt: altogether a very large improvement of 17pt (en-de, en-fr, at 60–61pt) to 25pt (en-ru, at 50pt) over [54].

Keung et al. [69] also evaluate the contribution of parallel sentences mined with their contextual-embedding-based method to unsupervised neural machine translation on the WMT 14 fr-en and WMT 16 de-en datasets. Their baseline model is XLM [60], which obtains much higher BLEU scores than [55]. They apply their method to mine 2.5 million parallel fr-en and de-en sentences pairs from Wikipedia dumps, and insert them in the unsupervised machine translation training process of [60] in a way similar to Hangya and Fraser [54]. This improves XML BLEU scores by 2.4pt to 3.5pt: en-de from 27.7 to 30.7, de-en from 34.5 to 37.3, en-fr from 36.7 to 40.2, and fr-en from 34.5 to 36.9.

The same method is used in a low-resource setting to collect from 100k to 400k pairs of English-Vietnamese parallel sentences from Wikipedia. These sentence pairs are added to the small parallel corpus of 133,000 English-Vietnamese sentence pairs of the IWSLT'15 dataset to train the 6+6-layer transformer model of [61]. This improves its BLEU score by 1.2pt from 32.8 to 34.0 (300,000 mined parallel sentence pairs). To create a very low resource context, they experiment with datasets 10 times smaller and the addition of 10k to 40k mined sentence pairs. In that setting, the improvements in BLEU scores are even larger, with an increase of 4–7pt from 20.7 to 27.7.

Sun et al. [52] also start from multilingual BERT to initialize multilingual word embeddings and sentence embeddings. They use a language discriminator to push the embeddings towards universal language-agnostic representations. These representations open the way to transfer learning: training a parallel sentence classifier on resource-rich language pairs such as English-French, and applying this classifier to low-resource pairs such as English-Esperanto and Chinese-Kazakh.

These methods are tested on the BUCC 2018 parallel sentence extraction dataset, which is the same as BUCC 2017. They obtain large improvements of 8–17pt in F1 scores over the unsupervised method of [69]: en-fr from 73.0 to 80.6, en-de from 74.9 to 86.9, en-ru fro 69.7 to 80.6, and en-zh from 60.1 to 77.3. The resulting F1 scores outperform the supervised method of [27] but remain 9–14pt below the supervised method of [14].

The low-resource setting is tested by training and testing the openNMT system with the parallel sentence pairs extracted from the Esperanto and Kazakh Wikipedias. This yields BLEU scores of 24.3 (en-eo) and 25.8 (zh-kk), against 18.5 and 21.6 for [54] and 20.2 and 22.8 for [69].

In conclusion, unsupervised parallel sentence extraction methods based upon unsupervised word or sentence embeddings have been steadily improving and are coming within smaller distance from supervised methods.

References

1. Adafre SF, de Rijke M (2006) Finding similar sentences across multiple languages in Wikipedia. In: Proceedings of the Workshop on NEW TEXT Wikis and blogs and other dynamic text sources. https://www.aclweb.org/anthology/W06-2810
2. Munteanu DS, Marcu D (2005) Improving machine translation performance by exploiting non-parallel corpora. Comput Linguist 31(4):477–504
3. Smith JR, Quirk C, Toutanova K (2010) Extracting parallel sentences from comparable corpora using document level alignment. In: Human language technologies: the 2010 annual conference of the North American chapter of the Association for Computational Linguistics, Los Angeles, CA, June 2010. Association for Computational Linguistics, pp 403–411. https://aclanthology.org/N10-1063
4. Kay M, Roscheisen M (1988) Text-translation alignment. Technical report, Xerox Palo Alto Research Center
5. Gale WA, Church KW (1993) A program for aligning sentences in bilingual corpora. Comput Linguist 19(1):75–102. https://aclanthology.org/J93-1004
6. Wu D (1994) Aligning a parallel English-Chinese corpus statistically with lexical criteria. In: Proceedings of the 32nd annual meeting on Association for Computational Linguistics (ACL '94), Stroudsburg, PA, USA. Association for Computational Linguistics, pp 80–87
7. Melamed ID (1999) Bitext maps and alignments via pattern recognition. Comput Linguist 25(1):107–130
8. Moore RC (2002) Fast and accurate sentence alignment of bilingual corpora. In: Machine translation: from research to real users, 5th conference of the association for machine translation in the Americas, Heidelberg, Germany. Springer, pp 135–244
9. Sennrich R, Volk M (2011) Iterative, MT-based sentence alignment of parallel texts. In: Proceedings of the 18th Nordic conference of computational linguistics (NODALIDA 2011), Riga, Latvia, May 2011. Northern European Association for Language Technology (NEALT), pp 175–182. https://aclanthology.org/W11-4624
10. Fluhr C, Bisson F, Elkateb F (2000) Parallel text alignment using crosslingual information retrieval techniques. Springer Netherlands, Dordrecht, pp 187–200. ISBN 978-94-017-2535-4. https://doi.org/10.1007/978-94-017-2535-4_9

11. Caswell I, Kreutzer J, Wang L, Wahab A, van Esch D, Ulzii-Orshikh N, Tapo A, Subramani N, Sokolov A, Sikasote C et al (2021) Quality at a glance: an audit of web-crawled multilingual datasets. arXiv:2103.12028
12. Agirre E, Banea C, Cer D, Diab M, Gonzalez-Agirre A, Mihalcea R, Rigau G, Wiebe J (2016) SemEval-2016 Task 1: semantic textual similarity, monolingual and cross-lingual evaluation. In: Proceedings of the 10th international workshop on semantic evaluation (SemEval-2016), San Diego, CA, June 2016. Association for Computational Linguistics, pp 497–511. https://doi.org/10.18653/v1/S16-1081
13. Zweigenbaum P, Sharoff S, Rapp R (2018) A multilingual dataset for evaluating parallel sentence extraction from comparable corpora. In: Proceedings of the eleventh international conference on language resources and evaluation (LREC 2018), Miyazaki, Japan, May 2018. European Language Resources Association (ELRA), pp 3828–3833. ISBN 979-10-95546-00-9. http://www.lrec-conf.org/proceedings/lrec2018/pdf/955.pdf
14. Artetxe M, Schwenk H (2019) Massively multilingual sentence embeddings for zero-shot cross-lingual transfer and beyond. Trans Assoc Comput Linguist 7:597–610
15. Hu J, Ruder S, Siddhant A, Neubig G, Firat O, Johnson M (2020) XTREME: a massively multilingual multi-task benchmark for evaluating cross-lingual generalisation. In: Daum H III, Singh A eds, Proceedings of the 37th international conference on machine learning, volume 119 of Proceedings of machine learning research, pp 4411–4421. PMLR, 13–18 Jul 2020. https://proceedings.mlr.press/v119/hu20b.html
16. Huang E, Socher R, Manning C, Ng A (2012) Improving word representations via global context and multiple word prototypes. In: Proceedings of the 50th annual meeting of the association for computational linguistics (Volume 1: Long Papers), Jeju Island, Korea, Jul 2012. Association for Computational Linguistics, pp 873–882. https://aclanthology.org/P12-1092
17. Huang J, Cai X, Church K (2020) Improving bilingual lexicon induction for low frequency words. In: Proceedings of the 2020 conference on empirical methods in natural language processing (EMNLP), Online, Nov 2020. Association for Computational Linguistics, pp 1310–1314. https://doi.org/10.18653/v1/2020.emnlp-main.100. https://aclanthology.org/2020.emnlp-main.100
18. Irvine A, Callison-Burch C (2016) End-to-end statistical machine translation with zero or small parallel texts. Nat Lang Eng 22(4):517–548. https://doi.org/10.1017/S1351324916000127
19. Irvine A, Callison-Burch C (2017) A comprehensive analysis of bilingual lexicon induction. Comput Linguist 43(2):273–310
20. Jakubina L, Langlais P (2016) BAD LUC@WMT 2016: a bilingual document alignment platform based on lucene. In: Proceedings of the first conference on machine translation: volume 2, shared task papers, Berlin, Germany, Aug 2016. Association for Computational Linguistics, pp 703–709. https://doi.org/10.18653/v1/W16-2370. https://aclanthology.org/W16-2370
21. Jantunen J (2002) Comparable corpora in translation studies: strengths and limitations. SKY J Linguist I5(43):105–117
22. Jegou H, Douze M, Schmid C (2011) Product quantization for nearest neighbor search. IEEE Trans Pattern Anal Mach Intell 33(1):117–128. ISSN 0162-8828. 10.1109/TPAMI.2010.57. https://doi.org/10.1109/TPAMI.2010.57
23. Johnson J, Douze M, Jégou H (2021) Billion-scale similarity search with GPUs. IEEE Trans Big Data 7(3):535–547. https://doi.org/10.1109/TBDATA.2019.2921572
24. Utiyama M, Isahara H (2003) Reliable measures for aligning Japanese-English news articles and sentences. In: Proceedings of the 41st annual meeting of the association for computational linguistics, Sapporo, Japan, Jul 2003. Association for Computational Linguistics, pp 72–79. https://doi.org/10.3115/1075096.1075106. https://aclanthology.org/P03-1010
25. Fung P, Cheung P (2004) Mining very-non-parallel corpora: parallel sentence and lexicon extraction via bootstrapping and EM. In: Proceedings of the 2004 conference on empirical methods

in natural language processing, Barcelona, Spain, Jul 2004. Association for Computational Linguistics, pp 57–63. https://aclanthology.org/W04-3208
26. Zhao B, Vogel S (2002) Adaptive parallel sentences mining from web bilingual news collection. In: Proceedings of the 2002 IEEE international conference on data mining, pp 745–748. https://doi.org/10.1109/ICDM.2002.1184044
27. Schwenk H (2018) Filtering and mining parallel data in a joint multilingual space. In: Proceedings of the ACL, Melbourne, Australia, Jul 2018. Association for Computational Linguistics, pp 228–234. https://doi.org/10.18653/v1/P18-2037. https://www.aclweb.org/anthology/P18-2037
28. Hangya V, Braune F, Kalasouskaya Y, Fraser A (2018) Unsupervised parallel sentence extraction from comparable corpora. In: Proceedings of the 15th international workshop on spoken language translation (IWSLT), pp 7–13. https://www.cis.uni-muenchen.de/~fraser/pubs/hangya_iwslt2018.pdf
29. Harris Z (1954) Distributional structure. Word 10(23):146–162
30. Hermann KM, Blunsom P (2014) Multilingual models for compositional distributed semantics. In: Proceedings of the 52nd annual meeting of the association for computational linguistics, Baltimore, Maryland, June 2014. Association for Computational Linguistics, pp 58–68. https://doi.org/10.3115/v1/P14-1006. https://aclanthology.org/P14-1006
31. Hoshen Y, Wolf L (2018) Non-adversarial unsupervised word translation. In: Proceedings of the 2018 conference on empirical methods in natural language processing, Brussels, Belgium, Oct 2018. Association for Computational Linguistics, pp 469–478. https://doi.org/10.18653/v1/D18-1043. https://aclanthology.org/D18-1043
32. Abdul-Rauf S, Schwenk H (2009) Exploiting comparable corpora with TER and TERp. In: Proceedings of the 2nd workshop on building and using comparable corpora: from parallel to non-parallel corpora (BUCC), Singapore, Aug 2009. Association for Computational Linguistics, pp 46–54. https://aclanthology.org/W09-3109
33. Etchegoyhen T, Azpeitia A (2016) Set-theoretic alignment for comparable corpora. In: Proceedings of the 54th annual meeting of the Association for Computational Linguistics (Volume 1: Long Papers), Berlin, Germany, Aug 2016. Association for Computational Linguistics, pp 2009–2018. https://doi.org/10.18653/v1/P16-1189. https://aclanthology.org/P16-1189
34. Brown PF, Cocke J, Della Pietra SA, Della Pietra VJ, Jelinek F, Lafferty JD, Mercer RL, Roossin PS (1990) A statistical approach to machine translation. Comput Linguist 16(2):79–85
35. Buck C, Koehn P (2016) Findings of the WMT 2016 bilingual document alignment shared task. In: Proceedings of the first conference on machine translation: volume 2, shared task papers, Berlin, Germany, Aug 2016. Association for Computational Linguistics, pp 554–563. https://doi.org/10.18653/v1/W16-2347. https://aclanthology.org/W16-2347
36. Cai X, Huang J, Bian Y, Church K (2021) Isotropy in the contextual embedding space: clusters and manifolds. In: International conference on learning representations
37. Grover J, Mitra P (2017) Bilingual word embeddings with bucketed CNN for parallel sentence extraction. In: Proceedings of ACL 2017, student research workshop, Vancouver, Canada, Jul 2017. Association for Computational Linguistics, pp 11–16. https://aclanthology.org/P17-3003
38. Schwenk H, Douze M (2017) Learning joint multilingual sentence representations with neural machine translation. In: Proceedings of the 2nd workshop on representation learning for NLP, Vancouver, Canada, Aug 2017. Association for Computational Linguistics, pp 157–167. https://doi.org/10.18653/v1/W17-2619. https://aclanthology.org/W17-2619
39. Artetxe M, Schwenk H (2019) Margin-based parallel corpus mining with multilingual sentence embeddings. In: Proceedings of the 57th annual meeting of the association for computational linguistics, Florence, Italy, Jul 2019. Association for Computational Linguistics, pp 3197–3203. https://doi.org/10.18653/v1/P19-1309. https://aclanthology.org/P19-1309

40. Grégoire F, Langlais P (2018) Extracting parallel sentences with bidirectional recurrent neural networks to improve machine translation. In: Proceedings of the 27th international conference on computational linguistics, Santa Fe, New Mexico, USA, Aug 2018. Association for Computational Linguistics, pp 1442–1453. https://aclanthology.org/C18-1122
41. Kiros R, Zhu Y, Salakhutdinov RR, Zemel R, Urtasun R, Torralba A, Fidler S (2015) Skip-thought vectors. In: Cortes C, Lawrence N, Lee D, Sugiyama M, Garnett R (eds), Advances in neural information processing systems, volume 28. Curran Associates, Inc. https://proceedings.neurips.cc/paper/2015/file/f442d33fa06832082290ad8544a8da27-Paper.pdf
42. Alexandre K, Ivan T, Binod B (2012) Inducing crosslingual distributed representations of words. In: Proceedings of the COLING, Mumbai, India
43. Koehn P (2005) Europarl: a parallel corpus for statistical machine translation. In: Proceedings of the MT Summit X
44. Koehn P (2010) Statistical machine translation. Cambridge University Press. ISBN 9780521874151. https://books.google.gr/books?id=4v_Cx1wIMLkC
45. Koehn P (2020) Neural machine translation. Cambridge University Press. ISBN 9781108497329. https://books.google.gr/books?id=mdDqygEACAAJ
46. Koppel M, Ordan N (2011) Translationese and its dialects. In: Proceedings of the 49th annual meeting of the association for computational linguistics: human language technologies, Portland, Oregon, USA, June 2011. Association for Computational Linguistics, pp 1318–1326. https://www.aclweb.org/anthology/P11-1132
47. McCann B, Bradbury J, Xiong C, Socher R (2017) Learned in translation: contextualized word vectors. arxiv:1708.00107
48. Artetxe M, Labaka G, Agirre E (2017) Learning bilingual word embeddings with (almost) no bilingual data. In: Proceedings of the ACL, Vancouver, pp 451–462
49. Lample G, Conneau A, Ranzato MA, Denoyer L, Jégou H (2018) Word translation without parallel data. In: Proceedings of the international conference on learning representations
50. Laroche A, Langlais P (2010) Revisiting context-based projection methods for term-translation spotting in comparable corpora. In: Proceedings of the 23rd international conference on computational linguistics (Coling 2010), Beijing, China, Aug 2010. Coling 2010 Organizing Committee, pp 617–625. https://aclanthology.org/C10-1070
51. Alvarez-Melis D, Jaakkola T (2018) Gromov-Wasserstein alignment of word embedding spaces. In: Proceedings of the 2018 conference on empirical methods in natural language processing, Brussels, Belgium, Oct 2018. Association for Computational Linguistics, pp 1881–1890. https://doi.org/10.18653/v1/D18-1214. https://aclanthology.org/D18-1214
52. Sun Y, Zhu S, Yifan F, Mi C (2021) Parallel sentences mining with transfer learning in an unsupervised setting. In: Proceedings of the 2021 conference of the North American chapter of the association for computational linguistics: student research workshop, Online, June 2021. Association for Computational Linguistics, pp 136–142. https://doi.org/10.18653/v1/2021.naacl-srw.17. https://aclanthology.org/2021.naacl-srw.17
53. Devlin J, Chang M-W, Lee K, Toutanova K (2019) BERT: pre-training of deep bidirectional transformers for language understanding. In: Proceedings of the 2019 conference of the North American chapter of the association for computational linguistics: human language technologies, volume 1 (Long and Short Papers), Minneapolis, Minnesota, June 2019. Association for Computational Linguistics, pp 4171–4186. https://aclanthology.org/N19-1423
54. Hangya V, Fraser A (2019) Unsupervised parallel sentence extraction with parallel segment detection helps machine translation. In: Proceedings of the ACL, Florence, Italy, Jul 2019. Association for Computational Linguistics, pp 1224–1234. https://doi.org/10.18653/v1/P19-1118. https://www.aclweb.org/anthology/P19-1118

55. Artetxe M, Labaka G, Agirre E, Cho K (2018) Unsupervised neural machine translation. In: 6th international conference on learning representations, ICLR 2018, Vancouver, BC, Canada, 3 Apr 30–May 2018, conference track proceedings. OpenReview.net, 2018. https://openreview.net/forum?id=Sy2ogebAW
56. Artetxe M, Labaka G, Agirre E (2019) An effective approach to unsupervised machine translation. arXiv:1902.01313
57. Aston G, Burnard L (1998) The BNC handbook: exploring the British national corpus with SARA. Edinburgh University Press, Edinburgh
58. Baayen H (2008) Analyzing linguistic data. Cambridge University Press, Cambridge
59. Bański P, Gozdawa-Gołębiowski R (2010) Foreign language examination corpus for l2-learning studies. In: Proceedings of the workshop on building and using comparable corpora, Malta
60. Conneau A, Lample G (2019) Cross-lingual language model pretraining. In: Conference on neural information processing systems, Vancouver, Canada, pp 7059–7069
61. Nguyen TQ, Salazar J (2019) Transformers without tears: improving the normalization of self-attention. In: 16th international workshop on spoken language translation, Hong Kong, Nov 2019. Zenodo. https://doi.org/10.5281/zenodo.3525484
62. Schwenk H, Chaudhary V, Sun S, Gong H, Guzmán F (2021) WikiMatrix: mining 135M parallel sentences in 1620 language pairs from Wikipedia. In: Proceedings of the 16th conference of the European chapter of the Association for Computational Linguistics: main volume, Online, Apr 2021a. Association for Computational Linguistics, pp 1351–1361. https://aclanthology.org/2021.eacl-main.115
63. Sarikaya R, Maskey S, Zhang R, Jan E-E, Wang D, Ramabhadran B, Roukos S (2009) Iterative sentence-pair extraction from quasi-parallel corpora for machine translation. In: Proceedings of the Interspeech 2009, pp 432–435. https://doi.org/10.21437/Interspeech.2009-156
64. Schwenk H, Wenzek G, Edunov S, Grave E, Joulin A, Fan A (2021) CCMatrix: mining billions of high-quality parallel sentences on the web. In: Proceedings of the 59th annual meeting of the association for computational linguistics and the 11th international joint conference on natural language processing (Volume 1: Long Papers), Online, Aug 2021. Association for Computational Linguistics, pp 6490–6500. https://aclanthology.org/2021.acl-long.507
65. Munteanu DS, Fraser A, Marcu D (2004) Improved machine translation performance via parallel sentence extraction from comparable corpora. In: Proceedings of the human language technology conference of the North American chapter of the association for computational linguistics: HLT-NAACL 2004, Boston, MA, USA, May 2004. Association for Computational Linguistics, pp 265–272. https://aclanthology.org/N04-1034
66. Stefănescu D, Ion R, Hunsicker S (2012) Hybrid parallel sentence mining from comparable corpora. In: Proceedings of the 16th annual conference of the European association for machine translation, Trento, Italy, May 2012. European Association for Machine Translation, pp 137–144. https://aclanthology.org/2012.eamt-1.37
67. Sánchez-Cartagena VM, Bañón M, Ortiz-Rojas S, Ramírez G (2018) Prompsit's submission to WMT 2018 parallel corpus filtering shared task. In: Proceedings of the third conference on machine translation: shared task papers, Belgium, Brussels, Oct 2018. Association for Computational Linguistics, pp 955–962. https://doi.org/10.18653/v1/W18-6488. https://aclanthology.org/W18-6488
68. Fan A, Bhosale S, Schwenk H, Ma Z, El-Kishky A, Goyal S, Baines M, Celebi O, Wenzek G, Chaudhary V, Goyal N, Birch T, Liptchinsky V, Edunov S, Auli M, Joulin A (2021) Beyond English-centric multilingual machine translation. J Mach Learn Res 22(107):1–48. http://jmlr.org/papers/v22/20-1307.html
69. Keung P, Salazar J, Lu Y, Smith NA (2020) Unsupervised bitext mining and translation via self-trained contextual embeddings. Trans Assoc Comput Linguist 8:828–841

70. España-Bonet C, Varga ÁC, Barrón-Cedeño A, van Genabith J (2017) An empirical analysis of NMT-derived interlingual embeddings and their use in parallel sentence identification. IEEE J Sel Top Signal Process 11(8):1340–1350. https://doi.org/10.1109/JSTSP.2017.2764273. https://doi.org/10.1109/JSTSP.2017.2764273
71. Feng F, Yang Y, Cer D, Arivazhagan N, Wang W (2022) Language-agnostic BERT sentence embedding. In: Proceedings of the 60th annual meeting of the association for computational linguistics (Volume 1: Long Papers), Dublin, Ireland, May 2022. Association for Computational Linguistics, pp 878–891. https://doi.org/10.18653/v1/2022.acl-long.62. https://aclanthology.org/2022.acl-long.62

Induction of Bilingual Dictionaries 5

5.1 Setting the Task

The aim of the Bilingual Lexicon Induction (BLI) task is to produce a bilingual lexicon using a pair of comparable corpora and either a small set of seed translations (a supervised setting) or no seeds at all (an unsupervised setting). A traditional bilingual dictionary usually offers a structure of senses and conditions for their translations, as well as POS tags for disambiguation. In contrast to this task, building bilingual lexicons as the aim for the BLI task involves a number of simplifications. The lexicons are most often represented via a simple list of word pairs, see the left side of Table 5.1 for an example. In this simplification, senses, subsenses and conditions for their translation are not indicated, while multiword translations are often omitted or listed through single words, for example, instead of listing a proper translation of *kindly* into French as *plein de gentillesse* a bilingual lexicon might list *gentillesse*, which is a noun meaning 'kindness' in contrast to the original expressions *kindly* and *plein de gentillesse*, both of which function as adjectives in this sense. Also the morphological variants are often repeated for each form, for example, *gentil* masculine versus *gentille* feminine in the example of Table 5.1. So the BLI task is a simplification, but it is still useful for many cross-lingual applications, such as information retrieval, text classification or lexicon enrichment for Machine Translation.

Typically, BLI methods return ranked lists of candidate translations for each source word. The traditional intrinsic measures for evaluating such lists are Precision@K, i.e., the proportion of correct translations at the top K positions in a list of candidates with the proportions counted over the gold standard list. The commonly used values of K are 1, 5 and 10, see [1]. Precision@1 is a natural measure for assessing the top suggested translation candidate as either correct or wrong. However, this measure can be unnecessarily strict, as it does not take into account inherent translation variation when (1) more than one translation

S. Sharoff et al., *Building and Using Comparable Corpora for Multilingual Natural Language Processing*, Synthesis Lectures on Human Language Technologies,
https://doi.org/10.1007/978-3-031-31384-4_5

Table 5.1 Comparing a bilingual lexicon to a dictionary

Lexicon	Dictionary	
kindly gentil	**kindly**.adj	1. (person) gentil
kindly gentille		2. (face) sympathique
kindly sympathique		3. (voice) plein de gentillesse
kindly gentillesse	**kindly**.adv	1. (in a nice way) avec gentillesse
kindly gentiment		2. (obligingly) gentiment
kindly apprécier		3. *to take kindly to* apprécier

option is feasible and (2) this is not reflected in the gold standard list. In many evaluation scenarios, Precision@10 is chosen as a measure, which aims at replicating human judgement with the possibility of selecting something useful in the output of 10 options at the top of a list. However, choosing P@10 has the problem of not distinguishing how the candidates are ranked *within* the top 10 candidates. Yet another measure of intrinsic BLI performance is Mean Average Precision (MAP), which takes into account the position of each correct answer. At the same time, MAP has a problem of utility for downstream BLI tasks, for which the right candidates in positions 11 or 100 are likely to be equally useless, while they are distinguished by MAP. For example, if a BLI method suggests a ranked list of *doucement, gentiment* for *kindly*, and this method is checked against the gold standard list in Table 5.1 (*doucement* can be a valid translation for *kindly* as ADV in some contexts), this BLI method will score 0 according to P@1 for this gold standard, while scoring 1 according to P@10 and 0.5 according to MAP (since *doucement* is not in the gold standard list used for testing).

5.2 Bilingual Lexicon Induction From Parallel Corpora

For parallel corpora, i.e. collections of texts where each source language text is paired with its target language translation, the solution to this problem stems to [2] by parallelizing the sentences of the two texts via sentence alignment, and then finding the matches between corresponding words of the aligned sentences via word alignment. Following the methods discussed in Chap. 4, it is possible to turn comparable corpora into collections of individual sentences. However, in this case we need to deal with the impact of noise when parallel sentences are extracted from comparable resources.

With truly parallel corpora used, the results of bilingual lexicon induction using parallel corpora are usually rather good. Tables 5.2 and 5.3 show some sample results as extracted from the German-English and French-English parts of the Europarl corpus [3] after a sentence alignment conducted with the length-based [4] aligner and after conducting word alignment using the Giza++ tool [5]. Note that the unfiltered versions of the dictionaries look noisy on first glance as word alignment is a sophisticated task leading to many errors.

Table 5.2 Consecutive lines from a German–English lexicon as derived from Europarl

Translation rate	Frequency of source	Frequency of translation	Source word	Translation
43.01	93	40	anfällig	vulnerable
15.79	19	3	anfällige	sensitive
52.63	19	10	anfällige	vulnerable
13.64	22	3	anfälligen	fragile
36.36	22	8	anfälligen	vulnerable
13.04	23	3	anfälliger	fragile
43.48	23	10	anfälliger	more vulnerable
41.18	17	7	anfälligsten	most vulnerable
25.86	58	15	anfänglich	initially
60.00	30	18	anfängliche	initial
61.02	59	36	anfänglichen	initial
10.71	28	3	anfängt	begins
14.29	28	4	anfängt	start
10.71	28	3	anfängt	starting
17.86	28	5	anfängt	starts

If there is one line per word equation, an unreliable word equation which occurs only once will take as much space in the list of results as a highly reliable word equation occurring many times. However, if in the results spurious word alignments are filtered out, the lists look rather good. In Tables 5.2 and 5.3 the following thresholds have been used for filtering:

- Translation rate: ≥10%
- Absolute frequency of translation: ≥3

With these parameters, the dictionaries have a size in the order of 50,000 entries for each of the two language pairs. An important advantage of these automatically generated raw dictionaries in comparison to conventional dictionaries is that, as shown in the tables, for each pair of a word and its translation absolute and relative frequencies are known which is very important for purposes such as statistical machine translation.

Table 5.3 Consecutive lines from a French–English lexicon as derived from Europarl

Translation rate	Frequency of source	Frequency of translation	Source word	Translation
53.46	578	309	absurde	absurd
55.07	69	38	absurdes	absurd
17.98	89	16	absurdité	absurd
30.34	89	27	absurdité	absurdity
16.85	89	15	absurdité	nonsense
23.33	30	7	absurdités	absurdities
23.33	30	7	absurdités	nonsense
49.23	906	446	abus	abuse
19.98	906	181	abus	abuses
47.62	42	20	abusent	abuse
26.47	102	27	abuser	abuse
14.58	48	7	abusif	abuse
14.58	48	7	abusif	abused
12.50	48	6	abusif	misuse
15.73	89	14	abusive	abuse
10.11	89	9	abusive	improper
11.24	89	10	abusive	misuse
17.65	34	6	abusivement	abused
17.65	34	6	abusivement	misused
27.59	29	8	abusives	abusive
13.79	29	4	abusives	improper
38.24	34	13	abusé	abused
30.00	10	3	abusées	abused
33.33	30	10	abusés	abused
10.00	30	3	abusés	misled

5.3 Matching Contexts

Unrelated comparable corpora are less likely to produce a substantial number of parallel sentences. The use of translations is also related to translationese. This led to the popularity of Vector Space Models (VSMs). The vectors are typically based on the counts of either word-to-document co-occurrences [6] (also known as Latent Semantic Analysis) or word-to-word co-occurrences [7], see more examples in Chap. 2. The application of the methods to the BLI task is based on the assumption that if two words in a corpus of language A co-occur more often than expected by chance, then the translations of these two words should also co-occur more often than expected in a comparable corpus of language B [8]. For example, the words *teacher* and *school* usually co-occur more often than expected in corpora of English. Therefore, the German translations of *teacher* and *school*, namely *Lehrer* and *Schule*, should also co-occur more often than expected in German corpora.

Most of the time the vectors undergo a dimensionality reduction procedure, such as Singular Value Decomposition (SVD) or Random Semantic Indexing because the number of dimensions for the keywords or documents measures in many thousands, see the examples in Chap. 2.

Reference [9] suggested how this can be used for the BLI task via extending an existing basic dictionary, which can typically comprise a few thousand common words, often referred to as a *seed dictionary*. To give an example, Table 5.4 shows a small extract of such a seed dictionary.

Table 5.5 shows an extract of a large co-occurrence matrix of the target language. We assume that it has been derived by counting word co-occurrences in a large corpus and by applying some association measure, such as the log-likelihood ratio, on the co-occurrence counts. All matrix positions showing significant co-occurrences (e.g. above a chosen threshold of the log-likelihood score) are marked with a dot. Note that this matrix is not quadratic as its rows correspond to all words considered well represented in the corpus (e.g. above a corpus frequency of 20), but its columns are only the words occurring on the target language side of the seed dictionary. The reason is that, as we will see below, only these words can be utilized in the classic approach and therefore all other columns can be abandoned.

Given the seed dictionary and the target language co-occurrence matrix, if we now want to determine the English translation of a particular word in the source language, let's say of the German word *Schüler* which means *pupil*, we need to determine the German co-occurrence vector of *Schüler*. This can be done in the same way as we did for computing the English co-occurrence matrix, only instead of the English corpus we use a comparable German corpus. Also, we only need to compute the row for *Schüler*, rather than the full matrix. Let's assume that this process led to the co-occurrence vector as shown in Table 5.3. Note that in reality this vector would be much longer as it is supposed to contain

Table 5.4 Small extract of a seed lexicon

German	English
blau	blue
grün	green
Farbe	color
Junge	boy
Klasse	class
Schule	school

Table 5.5 Co-occurrence matrix

	blue	boy	class	color	green	school
black	•			•	•	
blue				•	•	
boy						•
class						•
color	•				•	
green	•			•		
ocean	•					
pupil		•	•			•
school			•			
teacher			•			•
white	•			•	•	

Table 5.6 German co-occurrence vector of *Schüler*

	blau	Farbe	Ferien	grün	Junge	Klasse	Note	Pflanze	Schule
Schüler			•		•	•	•		•

Table 5.7 Translated co-occurrence vector of/Schüler/showing unknown translations

	blue	color	???	green	boy	class	???	???	school
Schüler			•		•	•	•		•

the co-occurrence of *Schüler* with all German word that are well represented in the German corpus (e.g. above corpus frequency 20).

In the next step we use our seed dictionary to translate the words corresponding to the columns of Table 5.6 into English. As the seed dictionary is not very comprehensive, it will usually only be able to translate common words. All other words cannot be translated which leads us to Table 5.7 where unknown translations are indicated by "/???/". As the columns with unknown translations are of no use to us, we simply eliminate them, leading us to Table 5.8. As the last step in vector transformation, we finally have to sort the vector entries alphabetically according to the column headings as shown in Table 5.9.

The resulting vector is now ready to be compared line by line to all rows of the co-occurrence matrix shown in Table 5.5. If we assume that identical vector positions add 1 to a comparative score, whereas different positions add 0, we get the results as shown in Table 5.10. As can be seen, the correct translation of *Schüler*, namely pupil, gets the highest score of 6. Next, with a score of 5, comes *teacher*, i.e. another type of person strongly

Table 5.8 Translated co-occurrence vector of/Schüler/with columns of unknown translations eliminated

	blue	color	green	boy	class	school
Schüler				•	•	•

Table 5.9 Translated co-occurrence vector of *Schüler* after sorting

	blue	boy	class	color	green	school
Schüler		•	•			•

Table 5.10 Resulting scores (last column) when the vector of *Schüler* is compared to all rows of the co-occurrence matrix

German	blau	Junge	Klasse	Farbe	grün	Schule	
English	blue	boy	class	color	green	school	
Schüler		•	•			•	
	blue	boy	class	color	green	school	
black	•			•	•		→ 0
blue				•	•		→ 1
boy						•	→ 4
class						•	→ 4
color	•				•		→ 1
green	•			•			→ 1
ocean	•						→ 2
pupil		•	•			•	→ 6
school			•				→ 4
teacher			•			•	→ 5
white	•			•	•		→ 0

associated to *school*. In this walkthrough example we used binary values in the vectors and co-occurrence matrices. However, as through binarisation some information is lost, practical systems will typically stay with the floating point values resulting from applying an association measure, and in this case for vector comparison often the cosine similarity measure will be used.

To give an impression of the outcome of the method, Fig. 5.1 lists some sample results for a few common words. As the examples of *Butter* (butter) and *Häuschen* (cottage) show, sometimes there occurs a confusion between the correct translation and closely related words. This can be expected and is in analogy to our walkthrough example shown in Table 5.10

test word	expected translation & rank		computed translations (top 5)
Arzt	doctor	1	doctor patients patient hospital NHS
Baby	baby	1	baby child mother daughter father
Bett	bed	1	bed morning room sleep day
Bibel	Bible	1	Bible book religion read text
Brot	bread	1	bread cheese meat food butter
Butter	butter	2	cheese butter carton bread sugar
Erde	earth	1	earth surface planet sea Earth
Frau	woman	2	man woman boy friend wife
Häuschen	cottage	2	bungalow cottage house hut village
Junge	boy	1	boy lady girl friend man
Käse	cheese	1	cheese sausage meat bacon bread
Kind	child	1	child daughter son father mother
Kohl	cabbage	17074	Major Kohl Thatcher Gorbachev Bush

Fig. 5.1 Results of the classic method of BLI for some common words

where, as the translation to *Schüler*, the related word *teacher* came pretty close to the correct translation *pupil*. Depending on the test sets used, the percentage of correctly translated words can vary widely e.g. between 30% and 70% depending on how common the words to be translated are, and on how well the target language corpus covers occurrences of its translation. In the example of Table 4.7 the system had no chance to come up with the correct translation of *Kohl* which is supposed to be *cabbage*, as in the underlying corpora the surname of the former German chancellor *Helmut Kohl* dominated.

These methods can involve a large number of hyper-parameters for building and aligning the vector spaces across the languages, such as the window size (from only considering the nearest neighbours to expanding the context to dozens of words), the measure for the collocation strength (for example, PMI or the log-likelihood (LL) score), the similarity distance, the size of the seed lexicon. These parameters have been evaluated by [10] on small English–French corpora extracted from Wikipedia. The surprising finding is that more commonly used measures, such as LL and cosine, are often less efficient in comparison to the log-odds measure for the collocation strength and the Dice score for the similarity distance.

A drawback of the classic method for BLI as described above is that it requires a seed lexicon. However, given comparable corpora with specific properties, it can sometimes be possible to do without. Imagine, for example, that we have large collections of English and German newspaper articles covering many years, and that we know the publication dates of all articles. Now imagine that, instead of using a co-occurrence matrix, we worked with a term/document matrix where in the rows remain the words, but where in the columns we have publication dates, i.e. a column represents all articles published on this particular day.

When now comparing row vectors between such an English and a German term/document matrix, we would notice that in the English matrix the word *Christmas* would strongly occur at dates around December each year, and likewise in German the word *Weihnachten* would show the same pattern. Thus by simple vector comparisons we could correctly conclude that *Weihnachten* is likely to be the translation of *Christmas*.

Once the BLI mappings for single words are known, this procedure has been applied to mining translations for multiword expressions by checking the n-gram translation candidates [11]. For example, *recreational fear* is an unusual expression as used in the context of why mass media focus on negative events,[1] so it is difficult to find a suitable translation which is not an awkward calque, such as *peur récréative*. However, it is possible to use the nearest neighbours for its individual words in the VSM space:

recreational → *educational, agricultural, leisure, entertainment*
fear → *desire, anxiety, emotion, interest, sense*

Their combinations suggest expressions which can be found in corpora, such as *leisure interest* as a suitable paraphrase for a specific context, which allows a more commonly used French equivalent, such as *intérêt des loisirs*.

In addition to matching keywords or their distributions via direct translations, small seed dictionaries provide the possibility to train discriminative models which are capable of combining a wide variety of features that individually provide only weak indications of translation equivalence [12]. Faruqui and Dyer [13] propose a simple method of using VSM spaces for two languages and a seed dictionary with the aim to transform the two spaces by means of Canonical Correlation Analysis (CCA). If we know from the seed dictionary that *giraffe, kindly, zebra* and *girafe, gentil, zèbre* are respective translations, and we know their vectors in the two VSM spaces, we can use the CCA procedure to maximise the correlations between the multidimensional variables in the two languages, which are considered to be translations. More specifically, if $\mathbf{E}$ is the matrix of vectors for the English seed dictionary, while $\mathbf{F}$ is the same for French, CCA outputs two matrices $\mathbf{V}$ and $\mathbf{W}$, which project the initial $\mathbf{E}$ and $\mathbf{F}$ into a new shared embedding space:

$$\mathbf{V}, \mathbf{W} = \text{argmax}_{\mathbf{V},\mathbf{W}} \rho(\mathbf{VE}, \mathbf{WF}) \tag{5.1}$$

More information about the CCA procedure (which internally involves running SVD) is available from [14].

Studies based on counts of context matches have been superseded in the 2010s by the rise of neural word embeddings, which are built with the objective of predicting a word from its context or predicting the context from a word. Embedding spaces stemming from predictions have been shown to be empirically better than the VSMs based on co-occurrence counts [15]. Nevertheless, there are inherent links between co-occurrence contexts and embeddings, since the process for building neural word embeddings also performs *implicit factorisation*

[1] https://www.theguardian.com/world/2004/mar/21/terrorism.comment1.

of the matrix of co-occurrence counts, see Sect. 10.4.3 in [16]. The methods for aligning embedding spaces across the languages rely on the same principles of transformation of VSM spaces, but historically new alignment studies have been based on embeddings, so this is covered below in Sect. 5.5.

5.4 Geometric Properties of Word Embedding Spaces

Neural embedding methods not only lead to empirically better results, they also exhibit the possibility to capture semantic regularities by applying linear algebra transformations to vectors to obtain their nearest neighbors [17]:

$$vec(king) - vec(man) + vec(woman) \rightarrow vec(queen) \tag{5.2}$$

$$vec(Paris) - vec(France) + vec(Germany) \rightarrow vec(Berlin) \tag{5.3}$$

$$vec(Paris) - vec(France) + vec(Italy) \rightarrow vec(Rome) \tag{5.4}$$

$$vec(sushi) - vec(Japan) + vec(Germany) \rightarrow vec(bratwurst) \tag{5.5}$$

so that the nearest vector for an operation of analogical reasoning over vectors makes sense from the viewpoint of their meanings, as *Paris* to *France* is the same as what *Berlin* is to *Germany*. It has been also shown that embedding spaces can amplify biases inherent in their training data, for example, if the embedding vectors are trained on Wikipedia, the female counterpart of a doctor is predicted as a nurse [18]

$$vec(doctor) - vec(man) + vec(woman) \rightarrow vec(nurse) \tag{5.6}$$

as stemming from far more male doctors and female nurses listed in the English Wikipedia in comparison to female doctors and male nurses. In the end, analogical reasoning with word embeddings is not only about the similarity of meanings, but also about expectations about these meanings as expressed by the authors of texts in the corpora used for a particular VSM.

The surprising properties of neural embeddings led to the rise of studies investigating their geometric properties and their implications for linking words across languages. In particular, three notions are relevant here: isotropy, isomorphism and hubness.

Isotropy is the case when vectors for words are more uniformly distributed in the vector space, as higher isotropy provides an easier way to distinguish between the vectors. Isotropy of a vector space can be measured as the average distance, for example, the cosine distance, between random vectors [19]. Anisotropy, as a deviation from the isotropic condition, will affect the shared aligned space used for the BLI task, when deviations from isotropy differ for different languages or for different semantic areas within a language.

Given that all operations with the vector spaces rely on the notion of distance, it is important to ensure that the embedding spaces for different languages are isomorphic, i.e.

they can be maped to each other while preserving the intra-language distances. The degree of isomorphism can be quantified in different ways. An efficient simple measure is the relational similarity (RS) metric [20], which is defined as the Pearson correlation coefficient ρ of the cosine distances for each element of word pairs $< w_D, w_R >$ for their respective languages. It has been demonstrated that isomorphism across languages is easier to achieve with larger corpora, while smaller corpora and greater typological differences between the languages lead to less isomorphic spaces. For example, the word embeddings from unrelated corpora for English and French are likely to be more isomorphic than the same case for English and Russian (all of those languages have sufficiently large corpora, but French is closer to English), while we can expect greater isomorphism between English and Russian embedding spaces than between English and Belarusian ones (Belarusian and Russian are closely related, but the corpora available for Belarusian are much smaller).

Hubness is a related phenomenon which occurs in many cases of feature spaces with a large number of dimensions, i.e., some vectors happen to be in close proximity to many other vectors [21]. This makes them more common choices in the lexical retrieval tasks leading to a larger number of errors. Formally, a word w can be linked to a set of words $\mathcal{N}_k(w)$ for which this word is within their k nearest neighbours. Words with the largest $|\mathcal{N}_k(w)|$ are (typically unwanted) hubs. Often such words have restricted context of their use, e.g., *troops* (183), *retreated* (176), *cavalry* (156) are such hubs in the FastText English space induced from Wikipedia (the numbers in brackets refer to their $|\mathcal{N}_{20}|$ hubness index, i.e., there are 183 words for which the word *troops* is in the list of their 20 closest neighbours), while the median $|\mathcal{N}_{20}|$ hubness index on the English Wikipedia is merely 5. Dinu et al. (2014) observe that hubness becomes more pronounced after linear transformation for mapping monolingual embedding spaces to a shared space. The reason is that the objective for building the transformation matrix **W** for mapping embedding spaces leads to lower variance of the transformed vectors, which in turn means that the vectors (on average) are closer to each other. They suggest a way of mitigating hubness by using Global Correction (GC), i.e., by downgrading the similarity ranks for the items proportionally to their hubness index. A similar technique has been also used as an objective for learning the transformation matrix **W** to minimise the impact of hubness [22].

So far we looked at the geometry of word embedding spaces, where a word is mapped to a single vector, i.e., this is the case of embedding spaces for word types. If a word can be used with several very different meanings, such as *bank*, this maps to a single embedding vector, parts of which are closer to other words about finances (for example, *fund*). Depending on the method for building the embedding space, such words can be the only neighbours of *bank* or other parts of its embedding vector can be closer to other words about geographical locations (for example, *coast*). The BLI task itself is inherently defined in this way by mapping word types to their translations, so that words with several possible translations either have a single entry or several entries, but without their contexts.

However, in recent years contextual embedding models, such as BERT [23] proved to be more successful in downstream applications in comparison to word-type embedding spaces.

The contextual embedding models can build different embedding vectors for a token when it is used in different contexts, thus offering a way of representing different subsenses, for example, the contextual embeddings for *table* in the following two sentences:

- *I put my glass on the kitchen* ***table***. → *J'ai posé mon verre sur la* ***table*** *de la cuisine.*
- *The* ***table*** *lists all the products.* → *Le* ***tableau*** *contient la liste de tous les produits.*

are likely to be different in the upper transformer layers of such models because the transformer layers integrate information from their contexts *kitchen table* versus *table lists*. From the viewpoint of relationship between English and French, the different vectors produced for *table* are likely to align with different translations.

Geometry of contextual embeddings is more complicated than that of word embeddings. The same token in different contexts occupies different positions in this space. It has been shown that the contextual embedding models can successfully encode many kinds of syntactic and semantic relations without explicit training for this task [24]. In particular, this can be tested by taking individual words in annotated sentences, for example, *glass* and *table* in *I put my glass on the kitchen table* and training a model which starts from their embedding vectors and predicts whether this word is used in the syntactic function of a direct object (*glass* in this example) or not (*table*). See more examples of predictions of syntactic and semantic relations by contextual embeddings in [24].

With respect to the BLI task, there is some evidence that different senses of the same word occupy specific regions in the embedding space of the contextual embedding models, for example, a semantic concordance annotated with WordNet word senses [25] can be consistently predicted with such models as ELMO [26] or BERT [27]. For example, the contextual embedding space makes it possible to obtain a simple nearest-neighbour classifier which can choose the sense of each word, such/*table*/$_1$ versus/*table*/$_2$ from the examples above. Even without additional fine-tuning, merely on the basis of searching for the nearest vector the sense-detection classifier achieved an F1 score of 0.71 [27], thus indicating that words used in separate senses are indeed separated in this space. This is a demonstration of the power of the contextual embeddings, but also this shows the limit of the traditional BLI test sets as they ignore differences in translating words in different senses.

There is a debate in the linguistic community concerning the anisotropic distribution of contextual embedding vectors. Some studies show that they occupy a narrow cone in the vector space, for example, the cosine similarity between randomly chosen English words in BERT ranges between 0.4 and 0.6 in the upper layers [28], i.e. unrelated words are similar to each other, thus leading to high anisotropy. At the same time, there are some observations that the high average cosine similarity does not necessarily imply the narrow cone [29, 30]. For example, Cai et al. [30] perform clustering of the embedding vectors for the entire Penn Treebank corpus for a number of contextual embedding models (BERT, D-BERT, ELMO, GPT and GPT-2) and show that for all of these models the token embeddings are located in several clusters and isotropy is nearly perfect *within* each cluster.

5.5 Alignment of Word Embeddings

Some methods for producing multilingual word embeddings used parallel corpora [31]. In this case the embedding representations are inherently parallel by the process of their construction. It has been also shown that the embeddings resulting from word-aligned corpora are also more isomorphic [32], thus leading to better quality of the BLI task. However, this approach is limited by the availability of truly parallel resources, see our discussion in the introduction to this book.

The requirement of using sentence- and word-aligned resources can be relaxed by using document-aligned corpora, such as Wikipedias [33]. However, comparable corpora natively aligned on the level of documents naturally lead to sentence-aligned resources (see Chap. 4 in this book), and thus to the inherently multilingual word embedding models.

The most common technique for alignment of word-level static embeddings is based on producing monolingual embeddings from unrelated comparable corpora and aligning them to produce an embedding space shared across the languages, either with a small seed dictionary or in an unsupervised manner. With respect to supervised approaches using seed dictionaries, it has been shown that simple linear transformations of the embedding vectors e_i and f_i in two languages are enough [34], see Fig. 2.1 in Sect. 1.3. The transformation matrix $\mathbf{W}$ can be computed by training on a seed dictionary through minimising the following loss function:

$$\min_{\mathbf{W}} \sum_{i \in \mathcal{D}} ||\mathbf{W}e_i - f_i||^2 \tag{5.7}$$

where $\mathcal{D}$ is a set of dictionary translations, usually from a small seed dictionary, for example, 100–2,000 entries have been shown to be sufficient [35]. Aligning the embedding spaces for more than a pair of languages is usually achieved via a pivot, when the set of embeddings e_i is aligned with f_i first, and embeddings from another language g_i can be aligned to the space of embeddings for either f_i or $\mathbf{W}e_i$ depending on the available seed dictionary.

Since the 2010s the majority of BLI approaches switched to using linear transformation of embeddings with main variations concerning different ways to obtain the transformation matrix $\mathbf{W}$, for example, via the stochastic gradient descent [34], canonical correlation analysis [13], multivariate regression [21] or orthogonal Procrustes transformation [36]. In the latter case the transformation matrix is built analytically as:

$$\mathbf{W} = \mathbf{V} \times \mathbf{U}^T \tag{5.8}$$

when $\mathbf{V}$ and $\mathbf{U}$ are the matrices from the SVD factorisation of $\mathbf{F} \times \mathbf{E}^T$, see [36] for justification and discussion.

Given that the presence of hubs has a negative impact on retrieval of the most suitable candidates, [37] suggested retrieving the candidates by replacing the cosine distance with Cross-Domain Similarity Local Scaling (CSLS), which regularises the cosine similarity with the mean density of the neighborhoods of the source vector v_e and the target vector v_f in the multilingual space:

$$CSLS(v_e, v_f) = 2cos(v_e, v_f) - knn(v_e) - knn(v_f) \tag{5.9}$$

knn is the mean of the cosine distance to its nearest neighbors.

All of the methods mentioned above build a transformation matrix as in (5.7) through minimising the Euclidean distance. A variation to these methods comes in the approach using Ranking-Based Optimization [22] which maximises a ranking-based objective between **F** and **E**. See [39] for a more comprehensive overview of approaches to building cross-lingual embedding spaces.

Unsupervised approaches to aligning embeddings started with an extension of the orthogonal Procrustes transformation method as in (5.8) via iterative bootstrap of aligned spaces [40]. In the first step, the seed dictionary is built from identical strings, such as numbers, punctuation marks or identical named entities. In the next step n best translation candidates are added to the seed dictionary and the procedure is repeated.

Later on, unsupervised approaches started using adversarial alignment methods, such as MUSE [37], which aim at matching the distribution of the embedding spaces in the two languages. This method actually extends the early distribution matching experiments from [8], namely that translations are likely to have similar distribution patterns. In the adversarial distribution matching setup, the objective consists in adversarial training of the generator and the discriminator. The discriminator is trained to distinguish whether a random sample comes from f_i or from $\mathbf{W}e_i$. In [37] the discriminator is implemented as a multilayer perceptron. The generator of better embedding spaces is the cross-lingual mapping procedure itself, which is trained to improve the transformation matrix $\mathbf{W}$, because a better transformation matrix should make the task of the discriminator more difficult. In the end, the loss function for the generator is the same as in inferring the transformation matrix, for example, (5.7), while for the discriminator, this is cross-entropy on a random sample of embeddings from the two languages:

$$\mathcal{L}_{cross} = \frac{1}{n}\sum \log P(\text{source} = 0|\mathbf{W}e_i) - \frac{1}{m}\sum \log P(\text{source} = 1|f_i) \tag{5.10}$$

A related method approaches the BLI task via Optimal Transport methods based on the Gromov-Wasserstein Alignment [41]. Given that the BLI task is aimed at finding an optimal mapping between possible translation equivalents, this can be seen as an Optimal Transport problem with the GWA algorithm used to find the best match between the distribution of **distances** of embeddings across two languages.

When completely unsupervised methods were introduced, they were shown to achieve parity with supervised methods without requiring seed dictionaries [37]. However, later research shows that unsupervised methods exhibit considerable variation in their accuracy depending on the quality of the initial embeddings, as well as on morphological richness in some of the languages [42]. This variation in accuracy has been shown to lead to unpredictable results, which are much less likely in the case of supervised methods, see also a discussion of evaluation methods below. These considerations led to explorations of semi-supervised methods, for example [43], which in addition to an adversarial loss as in (5.10) suggest adding a seed dictionary loss:

$$\mathcal{L}_{W|\mathcal{D}} = \frac{1}{|\mathcal{D}|} \sum_{i \in \mathcal{D}} s(\mathbf{W}e_i, f_i) \tag{5.11}$$

where s is a similarity function, such as the Euclidean distance or the cosine similarity. Adding a small seed dictionary leads to more reliable BLI results across the majority of language pairs according to [43].

Often words in the target translation pairs have similar orthography, especially in the case of closely related languages, for example, *zebra* in English versus *zèbre* in French. In addition to relying on just the embedding vectors, extra information from word forms can be used to compute the Weighted Levenshtein distance from the character-level alignment of the respective word forms [44]:

$$WLD(s_e, s_f) = \frac{\sum_{(c_e, c_f) \in al(s_e, s_f)} (1 - p(c_f|c_e))}{\max(len(s_e), len(s_f))} \tag{5.12}$$

where s_e and s_f are words in the two languages, al is a set of their character-level alignments from an alignment model, for example [45], $p(c_f|c_e)$ is the probability of aligning c_e with c_f. Knowledge about the similarity of word forms can be used to improve the quality of the shared embedding space in iterations:

$$score(s_e, s_f) = \alpha \cos(e, f) + (1 - \alpha)(1 - WLD(s_e, s_f)) \tag{5.13}$$

where e and f are vectors for respectively s_e and s_f in the shared cross-lingual embedding space, while α is the relative weight of the cosine similarity with respect to WLD. As shown in [44] for under-resourced language pairs this can lead to substantial improvements, such as increasing P@1 from 0.43 to 0.84 for the Slovenian-Croatian pair. For slightly less related languages, such as the Slovenian-Polish pair, there are still improvements, though not as substantial, P@1 from 0.58 to 0.75.

Another possible issue in using the embedding spaces concerns problems with aligning relatively rare forms for much more frequent lemmas, for example, while *toser* ('to cough') is reasonably frequent in Spanish, its imperative forms are very rare, which leads to their poor quality embeddings and poor alignment with *to cough* in English, even though the forms and their translations are regular and predictable. Integration of inflectional morphology into the alignment and dictionary induction process can lead to considerable improvements [46].

A related problem concerns retrieval of low frequency words in general [47], since retrieval of low-frequency words is disproportionately affected by hubness, because they tend to have noisier embeddings (such words have fewer context for converging their random initialisations to more meaningful embeddings). This leads to the case when noisier embeddings are more affected by words with greater hubness, which 'steal' their nearest neighbours and decrease the accuracy of predictions. A solution suggested by [47] concerns a method of Hubless Nearest Neighbour (HNN) search, which adds the assignment matrix **P** such that $P_{i,j}$ is the probability of assigning f_j as the translation of e_i in the seed dictionary. Such translation matrices are available as a by-product of building Statistical Machine Translation [5], and we usually do not have this information for the seed dictionaries of comparable corpora. However, if the transformation matrix **W** has been computed by one of the methods discussed above, the assignment matrix can be approximated by minimising:

$$\sum_{i,j} P_{i,j} cos(\mathbf{W}e_i, f_i) - \epsilon H(\mathbf{P}) \tag{5.14}$$

where $H(\mathbf{P}) = -\sum P_{i,j} \log P_{i,j}$ is entropy of **P**. It has been shown that the use of HNN can lead to considerable improvements for lower frequency words, for example, raising P@1 from 0.63 to 0.80 in the lowest frequency bins for the French-English pair [47].

5.6 Alignment of Contextual Embeddings

Contextual embeddings produced by pre-trained transformers, such as multilingual BERT, became aligned in the process of their training in a way similar to unsupervised alignment of word embeddings because their training requires sharing the weights across the languages for each layer, so the weights become similar for processing semantically similar sentences across all of the languages [48]. Alignment of the embedding representation spaces via weight sharing can outperform the case of adding sentence-aligned parallel corpora to the training set to obtain an explicit Translation Language Model [49]. Later experiments showed some advantages of having explicit translation information coming from parallel resources [50], but the advantages are minimal, since the amount of truly parallel data does not scale to hundreds of languages used in multilingual pre-trained models, such as BERT or XLM-Roberta.

There are two aspects which make the BLI task slightly different in the case of contextual embedding models. First, they need to reduce their lexicons by using a Byte Pair Encoding (BPE) mechanism [51], in which less frequent words are split into a sequence of BPE tokens, each of which is a more frequent word or a subword. The process for building a BPE lexicon on a corpus aims at maximising the number of tokens covered by a lexicon of a fixed size. In the end, typical BPE tokenisers split less frequent tokens, for example:

English *characteristically* → *characteristic ##ally*
French *typiquement* → *typique ##ment*

This splitting necessitates extra efforts for applying the embedding models to a seed dictionary and obtaining the results, as the predicted translation for *characteristic* can be *typique*, but *##ally* does not necessarily predicts *##ment*. The BPE space is also shared across the languages of multilingual pre-trained models, so that the same BPE toke *die* represents the English words *to die* and *a die* in their various senses, as well as determiners in Afrikaans, Dutch and German (in these languages *die* is cognate to *the* in English). The weights from the pre-trained models help in inferring different embeddings corresponding to different contexts (*I shall die* vs *a punch and die set* vs *die Frau*, 'the woman'), but the result of BPE tokenisation is not always optimal, so considerable post-processing efforts are needed for re-alignment of the output of BPE tokenisation [52].

Second, contextual embedding models do not operate on the level of lexical units, *characteristic* or *table*, but on the level of tokens used in their context, so that those models for these tokens in such contexts as *the kitchen table* versus *the table of products* produce different embeddings at the level above the initial lookup table. This behaviour is the desired outcome of using such models, but this makes it more difficult to map individual contexts of lexical items to their dictionary definitions. Nevertheless, the volatile space of contextual embeddings can be the source of studying multilingual lexical semantics by taking into account how these embeddings predict dictionary properties, such as synonymy, antonymy, hypernymy, etc. [53].

5.7 Evaluation

5.7.1 Evaluation Experiments for BLI

Evaluation is at the heart of modern research in computational linguistics. With many different methods aimed at producing bilingual lexicons it is important to compare their contributions using shared datasets. It is also important to keep in mind that the bilingual lexicons are themselves a tool for other tasks, such as document classification, information retrieval or Machine Translation, so one can consider intrinsic evaluation scenarios, when the tools are tested by comparing them to gold standard lexicons, and extrinsic evaluation on specific downstream tasks. The extrinsic evaluation scenarios are discussed in Chap. 7. Here we describe examples of intrinsic evaluations.

Reference [1] tested the contribution of various bilingual embedding models with varying degrees of cross-lingual supervision, from training multilingual word embedding models using word-aligned or sentence-aligned parallel resources to aligning monolingual embeddings coming from unrelated comparable corpora. Both intrinsic and extrinsic scenarios have been tested. The methods available at the time of that study (2016) showed that their ability to produce cross-lingual embeddings drops with the amount of supervision via parallel resources with models requiring word-aligned resources generally performing better than those requiring sentence-aligned resources, which are better than models using unrelated comparable corpora. However, specifically for the BLI task the unrelated comparable corpora with a seed dictionary can achieve P@1 comparable to models using fully parallel resources.

Most recently for the BLI task, [54] performed an extensive comparison of supervised and unsupervised methods using only comparable corpora (Wikipedia dumps) on the same dataset over various seed dictionaries and language pairs. Table 5.11 lists some of their results comparing the supervised approaches, in particular, CCA [13], Procrustes [36], see also Eq. (5.8), iterative Procrustes with bootstrap of an initial seed dictionary, RCSLS [22], to the unsupervised ones, including Vecmap [40] (see Chap. 6 for a description), MUSE [37], ICP [55] and GWA [41].

Even though some studies claim that unsupervised approaches can achieve parity or can even surpass supervised ones [37], the large-scale evaluation in [54] does not support this claim. Similar performance is achievable between the supervised and unsupervised methods for closely related languages, such as English, French and German, for example, both MUSE and VECMAP without training data achieve better performance for English–French and English-German than the supervised methods in Table 5.11, which use a *1k* seed dictionary. However, some unsupervised methods fail to converge (MAP = 0.0) or their performance drops below that of the supervised ones even with a 1 k seed dictionary for less related language pairs, for example, when English or German need to be aligned with Croatian, Finnish, Russian or Turkish. The assumption of similarities in the distributions of embeddings, which is the corner stone of the unsupervised methods, does not hold for more distant languages. Also the supervised approaches with *bigger* seed dictionaries (3 k or 5 k) always outperform the unsupervised ones even for the more closely related languages. Interestingly, the increase in the seed dictionary size from 1 to 3 k always leads to significant improvements, while further increases are less helpful, and can even lead to slightly degraded precision when its size reaches 10–50 k, as shown in [35]. A possible explanation for the degraded performance is that the seed dictionaries have been extracted from existing parallel resources, while they have been ordered by the probability of translations, therefore larger seed dictionaries tend to contain more noise.

There is also a question concerning the quality of the gold standards used for testing. A commonly used resource for several evaluation studies is the MUSE dataset, which was created using a not fully specified "in-house Facebook translation tool" [37]. Closer analysis of this test set shows a number of problems, concerning both the kind of words it contains across all languages and the kinds of translation equivalents it has for specific language pairs. First, as discussed by [56], it contains a large number of proper nouns, which differ in their semantics by having an indexical function of pointing to a named object in comparison to content words which describe objects and actions with them. In the end, some methods can exhibit very different results on proper nouns in comparison to content words.

Table 5.11 BLI performance (MAP) for some methods and language pairs from [54]

Model	Dict	DE-FI	DE-FR	DE-HR	DE-RU	DE-TR	EN-DE	EN-FI	EN-FR	EN-HR	EN-IT	EN-RU	EN-TR
CCA	1K	0.241	0.422	0.206	0.308	0.153	0.458	0.259	0.582	0.218	0.538	0.336	0.218
CCA	3K	0.328	0.494	0.298	0.399	0.251	0.531	0.351	0.642	0.299	0.613	0.434	0.314
CCA	5K	0.353	0.509	0.318	0.411	0.280	0.542	0.383	0.652	0.325	0.624	0.454	0.327
PROC	1K	0.264	0.428	0.225	0.323	0.169	0.458	0.271	0.579	0.225	0.535	0.352	0.225
PROC	3K	0.340	0.499	0.308	0.413	0.260	0.532	0.365	0.642	0.307	0.611	0.449	0.320
PROC	5K	0.359	0.511	0.329	0.425	0.284	0.544	0.396	0.654	0.336	0.625	0.464	0.335
PROC-B	1K	0.354	0.511	0.306	0.392	0.250	0.521	0.360	0.633	0.296	0.605	0.419	0.301
PROC-B	3K	0.362	0.514	0.324	0.413	0.278	0.532	0.380	0.642	0.336	0.612	0.449	0.328
RCSLS	1K	0.288	0.459	0.262	0.361	0.201	0.501	0.306	0.612	0.267	0.565	0.401	0.275
RCSLS	3K	0.373	0.524	0.337	0.442	0.296	0.568	0.404	0.665	0.357	0.637	0.491	0.364
RCSLS	5K	0.395	0.536	0.359	0.458	0.324	0.580	0.438	0.675	0.375	0.652	0.510	0.386
VECMAP	–	0.302	0.505	0.300	0.322	0.253	0.521	0.292	0.626	0.268	0.600	0.323	0.288
MUSE	–	0.000	0.005	0.245	0.272	0.237	0.520	0.000	0.632	0.000	0.608	0.000	0.294
ICP	–	0.251	0.454	0.240	0.245	0.215	0.486	0.262	0.613	0.000	0.577	0.259	0.000
GWA	–	0.216	0.433	0.015	0.222	0.101	0.029	0.022	0.462	0.003	0.017	0.002	0.003

The overuse of proper nouns is only part of its large inventory of identical, often very infrequent, strings, for example, for the English–French pair:

bronson	bronson
dag	dag
fps	fps
hobbies	hobbies
kazimierz	kazimierz
snp	snp
xxiii	xxiii

which can skew predictions of some models.

As mentioned earlier, the issue of isomorphism affects all methods for producing aligned embedding spaces by a transformation matrix, since this assumes the possibility of linear mapping between the monolingual spaces. There are at least two considerations which can challenge this assumption. First, non-isomorphism of the embedding spaces can arise because of the typological differences between the respective languages, for example, because words with ambiguity differ across languages. For example, the vector for *table* in English needs to be in the neighborhood of both the furniture sense and the information object sense, while the French space is built with very different vectors for *table* and *tableau*, corresponding to the two respective senses, thus leading to possible distortions when they need to be aligned. Second, the geometry of monolingual embedding spaces crucially depends on the amount of raw text data available for training. In a recent experiment by [20] it has been shown that a smaller amount of available raw texts is the most important factor which can cause degenerate word vector spaces. For example, the Croatian Wikipedia is by far the smallest in the set studied in Table 5.11, thus leading to lower overall accuracy, see Column En-Hr, as well as a drop in the accuracy of unsupervised methods for Croatian.

5.7.2 Evaluation on Multilingual Termbanks

Terminology serves as one of the direct applications of the BLI task since multilingual termbanks support such activities as document authoring or translation with a number of such termbanks available, for example, from international companies,[2] the European Commission[3] or the United Nations.[4] New concepts, product names or descriptions appear regularly, so termbanks need to be updated with the updates covering a wide range of languages, far beyond the linguistic expertise of a single terminologist. In its design a termbank is closer

---

[2] https://www.microsoft.com/en-us/language.

[3] https://iate.europa.eu/.

[4] https://unterm.un.org/.

Table 5.12 IATE examples

ID	IATE-28492	IATE-29585	IATE-2245344
English	industrial design	pre-flight check; pre-flight inspection; preflight inspection	glassy sodium polyphosphate
French	dessin industriel	inspection pré-vol; contrôle pré-vol; vérification avant le vol	polyphosphates de sodium, vitreux
Polish	projektowanie przemysłowe; wzornictwo przemysłowe	Przegląd przedlotowy	szklisty polifosforan sodu
Slovenian	industrijsko oblikovanje	predpoletni pregled; predvzletni pregled	steklast natrijev polifosfat

to a BLI list than to a dictionary by having no structure of senses or conditions for their translations, compare the bilingual lexicon in Table 5.1 to three sample entries from EU's IATE in Table 5.12.

However, a termbank is still more complex than a simple BLI list, see a more elaborate description of termbanks in [57]. First, many terms are multi-word entries (70% of English terms in IATE are multiwords). Besides, the number of words in a term and in its translation need not be the same. Specifically, single-word terms may have multi-word terms as their translations and conversely. Second, the terms in a termbank need to be expressed in their dictionary form, even though they are detected in corpora in their word forms with morphological variation, for example, *pre-flight checks, sodium polyphosphates* in English, while variation is much more widespread in morphologically rich languages like Finnish or Polish, when more complex alterations might be need to convert corpus occurrences into entries *szklistego fosforanu sodu* (the genitive form) → *szklisty polifosforan sodu*. Third, termbanks normally contain examples of terminological variation, which are all linked to the same terminological concept, for example, *pre-flight check* versus *pre-flight inspection*, note also examples of orthographic variation in *pre-flight* versus *preflight*. In IATE, about 10% of the English entries contain variants.

Fourth, many terms are fairly infrequent even in large corpora. For example, there are nine occurrences of *sodium polyphosphate* in the English Wikipedia and no occurrences of *glassy sodium polyphosphate* (in the Wikipedia dump of 24 Nov 2019). At the same time, the structure of terms is often regular, which should simplify their recognition even from a small number of examples. There are numerous occurrences of *glassy carbon, glassy silica* and other glassy substances in Wikipedia, thus making *glassy sodium polyphosphate* a fairly likely term candidate. Similarly, single examples of *calcium phosphate* or *dihydrogen triphosphate* are enough for their recognition as terms and for deriving their translations, since they have expected term structures. Finally, a termbank usually contains terms in several languages all linked to a single concept, while the existing BLI testbeds, for example, those listed in Sect. 5.7.1, operate only with pairs of languages.

5.8 The BUCC 2020 Shared Task on Bilingual Dictionary Induction

In conjunction with the 13th Workshop on Building and Using Comparable Corpora (BUCC 2020) we organized a shared task on bilingual dictionary induction (BLI) from comparable corpora. For a number of pairs of common languages, including Chinese, English, French, German, Russian, and Spanish, the shared task participants were asked to automatically determine the translations of 6000 source language test words which were divided into three frequency ranges. Below, we give an overview on the shared task including the task definition, the data sets, the evaluation procedure and the results of the participating systems. A comprehensive description of the shared task, including system description papers of the shared task participants, can be found in the proceedings of the 13th BUCC workshop [58]. All resources can be downloaded from the shared task website.[5]

The motivation for the shared task was that quite a few research groups had been working on the BLI problem using a wide variety of approaches. However, as there was no standard way to measure the performance of the systems, the published results were not comparable and the pros and cons of the various approaches were not clear.

The shared task aimed at solving these problems by organizing a fair competition between systems. This was accomplished by providing corpora and bilingual datasets for a number of language pairs involving diverse language families, and by comparing the results using a common evaluation framework. However, as we anticipated that the corpora and datasets provided by us might not suit all needs, we divided the shared task into two tracks:

In the *closed track*, participants were required to only use the data provided by us. In this way equal conditions were ensured and, as the outcome of this track, the systems could be ranked according to the quality of their results.

In the *open track*, participants were free to use their own corpora and datasets. If possible, they were supposed to still use our evaluation data, but this was not mandatory. In their system description papers which were published in the workshop proceedings, participants were asked to describe why their systems were not suitable for the closed track, and discuss the pros and cons of their choices.

5.8.1 Resources

The corpora provided for the shared task were those of the Web-as-a-corpus initiative (WaCKy) for English, French, German and Russian [59]. The WaCKy corpora are cleaned-up web crawls of approximately 2 billion words per language except for Russian where the

[5] https://comparable.limsi.fr/bucc2020/bucc2020-task.html.

size is about 3 billion words. As for Spanish and Chinese no WaCKy corpora were available, we instead provided cleaned-up versions of the respective Wikipedias. However, we recommended to give preference to the WaCKy corpora if they were available for both languages of a language pair. The reason was that they seemed somewhat better suited for the dictionary induction task than Wikipedia as they were more comparable and provided a wider range of topics and genres. For the convenience of the participants, fastText embeddings [60] were also provided for all corpora.

As our datasets for training and testing, with the kind permission of Alexis Conneau, we used extracts from the MUSE dictionaries as described in [37]. This data is from a Facebook internal translation tool and contains word equations which appear to be sorted according to occurrence frequencies of the source language words. By listing sample bilingual word pairs for three frequency bands from the English to German list, Table 5.13 gives an impression of

Table 5.13 Sample data from three frequency bands of the English to German MUSE word translation data

High		Middle		Low	
English	German	English	German	English	German
the	die	parkes	parkes	egner	egner
the	der	compromising	gefährden	chotek	chotek
the	dem	compromising	kompromisse	jop	jop
the	den	dutchman	niederländer	bechtolsheim	bechtolsheim
the	das	dutchman	holländer	fabulae	fabulae
and	sowie	intangible	immateriellen	tiede	tiede
and	und	intangible	immaterielle	pillnitz	pillnitz
was	war	recife	recife	grzimek	grzimek
was	wurde	schlesinger	schlesinger	höfner	höfner
for	für	addendum	nachtrag	stammheim	stammheim
that	dass	addendum	ergänzung	bruhns	bruhns
that	das	addendum	addendum	cozio	cozio
with	mit	addendum	zusatz	draheim	draheim
from	vom	primus	primus	chillon	chillon
from	von	awakened	aufgeweckt	schauinsland	schauinsland
from	ab	awakened	geweckt	macassar	macassar
from	aus	awakened	erweckt	şenol	şenol
this	dieser	hysteria	hysterie	bijutsu	bijutsu
this	diese	plethora	fülle	torkel	torkel
this	das	hauling	schleppen	munificence	huld

what this looks like. As can be seen, a source language word can correspond to several target language words. Hereby there seems to be a limit of at most five target language words.

Rather than providing one large set of training data per language pair, we extracted bilingual word pairs relating to 2000 randomly selected source language words from each of three frequency ranges (high, mid and low frequency) of the respective MUSE dictionary. Thus our training sets consist of three separate parts relating to three different frequency ranges concerning the words on the source language side of the bilingual word pairs. Taking all three parts together, this gives 6000 source language words per language pair. As there can be several translations per source language word, the number of bilingual word pairs is larger than 6000.

Distinguishing between frequency ranges is of scientific interest as BLI algorithms typically work best for high and medium frequency words, and thus in scientific papers sometimes preferably the results for these are presented. However, in practice dictionaries of common words are often available already, and therefore the performance at low frequencies is usually of higher practical relevance.

Concerning the test datasets, they were produced in analogy to the training datasets. Only care was taken not to have any overlap concerning the lists of 6000 source language words between the two datasets.

5.8.2 Evaluation

For evaluation, participants of the closed track (for the open track this was optional) were asked to provide their results on the test data sets for the test words in each of the three frequency ranges. Hereby it was expected that for each source language word all its major translations were provided (where the definition of "major" was supposed to be inferred from the training data). The shared task organizers compared these translations to the translations as found in their (internal) gold standard data which was structurally identical to the training data. Only identical strings were considered correct, and the performance of the systems is determined by computing precision (P), recall (R), and F1-score, the latter being the official score for system ranking.

More precisely: The input to the system is a list of source language words, one per line. A system should return, for each input word one or more translations, in the form of tab-separated word pairs, each on its own line. For instance, in the English–French case, given the gold standard, test word list, and system output (tab-separated word pairs) shown in Fig. 5.2, the system would get credited for two true positives, one false positive, and two false negatives, hence

$$P = 2 / 3 = 0.67$$
$$R = 2 / 4 = 0.50$$
$$F1 = 0.57$$

Fig. 5.2 Sample gold standard, test word list and system output for the English–French case

gold standard	
bed	lit
bed	plumard
doctor	médecin
doctor	docteur

test set
bed
doctor

system output	
bed	lit
bed	futon
doctor	docteur

Table 5.14 shows the results of the participating teams for the closed track. The acronyms of the teams have the following meanings: CEN = Amrita School of Engineering, Center for Computational Engineering and Networking (CEN); LMU = LMU Munich, Center for Information and Language Processing; LS2N = Université de Nantes, TALN/LS2N; SW = Sida Wang. The results when distinguishing the three frequency bins are provided in Table 5.15.

Despite the fact that BLI is a challenging task, these results show that the best participating systems achieved consistently good results for a number of language pairs involving languages from related as well as from distant language families. Surprisingly, for some language pairs (e.g. de-en) the predictions for low frequency words were better than for words with high frequency which should be particularly useful when identifying the translations of technical terms.

While CEN and SW were limited to specific language directions, LMU and LS2N used generic and very similar approaches which were comparable across the language pairs. Both used Cross-Domain Similarity Local Scaling (CSLS) for retrieval instead of the traditional cosine similarity. The LMU approach relied on the ensemble to emphasise that the candidates are ranked high by multiple models [61]. The LS2N approach relied on concatenating BOW and Skip-Gram vectors and taking into account the relative frequency of the translation candidates [62].

Table 5.14 Overall results of the closed track

Language	Team	R	P	F1
de-en	CEN	15.3	5.2	7.7
	LMU	48.7	61.6	54.4
	LS2N	57.5	66.2	61.5
en-de	LMU	40.2	59.8	48.1
	LS2N	54.3	54.8	54.5
en-ru	LMU	33.9	37.8	35.8
	LS2N	32.6	38.7	35.4
ru-en	LMU	43.9	56.7	49.5
	LS2N	35.5	56.7	43.7
de-fr	LS2N	76.8	76.7	76.8
fr-de	LS2N	78.3	64.9	71.0
en-es	LS2N	63.8	61.4	62.6
es-en	LS2N	67.5	75.1	71.1
en-fr	LS2N	61.2	69.7	65.1
fr-en	LS2N	46.0	64.6	53.7
en-zh	SW	45.3	54.6	49.5
zh-en	SW	33.6	40.9	36.9

Table 5.15 Comparison of results by frequency bands for the closed track

Lang.	Team	High			Mid			Low		
		R	P	F1	R	P	F1	R	P	F1
de-en	CEN	9.0	4.0	5.5	15.0	4.9	7.4	27.0	6.6	10.6
	LMU	44.7	49.1	46.8	43.4	70.9	53.8	62.8	77.1	69.2
	LS2N	48.1	63.7	54.8	59.0	63.0	60.9	72.2	73.3	72.8
en-de	LMU	35.1	51.4	41.7	35.0	65.3	45.6	61.4	71.2	66.0
	LS2N	49.0	51.6	50.3	53.7	52.6	53.2	68.6	65.2	66.9
en-ru	LMU	38.0	41.0	39.4	30.7	39.1	34.4	29.5	30.3	29.9
	LS2N	47.7	36.5	41.3	34.3	25.7	29.4	21.4	22.5	22.0
ru-en	LMU	45.3	67.6	54.2	45.5	59.4	51.5	39.9	43.1	41.4
	LS2N	49.3	59.0	53.7	38.3	56.0	45.5	13.2	48.8	20.8

5.9 The BUCC 2022 Shared Task on Bilingual Terminology Extraction

5.9.1 Specifications of the Task

In the wake of the BUCC 2020 shared task on bilingual dictionary induction, the BUCC 2022 shared task [63] aimed to test methods that identify pairs of terms that are translations of each other in comparable corpora. The first objective of this task as compared to BUCC 2020 was to test the ability of bilingual dictionary induction methods to detect multi-word term translations. Its design also emphasized a gold standard created from the contents of the corpus rather than from a pre-existing bilingual dictionary.

More specifically, we made the following design choices:

Multiword entries Both single and multiword terms should be included in the gold standard.

Frequency The gold standard terms can vary in their corpus frequencies with some terms from the gold standard occurring only once in the respective comparable corpora for each language. The gold standard should be as complete as possible by design, and may therefore contain terms with few occurrences.

Domains A sample of two specialised domains has been used, the Wind Energy domain from the TTC project [64] and the medical domain;

Languages As in other BUCC shared tasks, this task intends to cover a variety of languages (English, Chinese, French and Russian) which provide variation in terms of orthography and linguistic typology; however, only the English–French pair was provided in 2022.

Task specification Participants were given:

– A set of in-domain comparable corpora C_i in languages L_i

– A list of terms D_i found in C_i for each language L_i

• A system must return an ordered list of pairs of terms in (D_1, D_2) that are translations of each other

• Evaluation was performed through Average Precision against the gold standard dictionary

The following principles were used to create pairs of comparable corpora and associated gold standard bilingual terminologies for this shared task.

Extracting comparable sub-corpora from a large parallel collection We started from a very large, diverse collection of sentence-aligned, parallel texts (CC-Align) available in a large number of language pairs. Having parallel texts made it possible to collect gold-standard term translations with a very large coverage of this specific corpus. Having a large collection made it possible to sample specialized sub-corpora on multiple topics.

Sampling thematic corpora from a general collection We extracted thematic corpora from the initial collection by defining sets of topic-oriented keywords in one of the languages, say L_1, and retrieving sentences that contained these keywords in C_1, creating the sub-corpus C_1^j.

Aligned sub-corpora C_i^j for $i \neq 1$ were then obtained by following the sentence alignments. This provided pairs of parallel sub-corpora (C_i^j, C_k^j).

Performing automatic term extraction from the parallel corpora We applied an automatic term extraction (ATE) method to each corpus C_i and collected all terms, including single-word terms and multi-word terms. This provided in turn lists of monolingual terms in each sub-corpus in each of the considered languages. We used TTC Term Suite [65, 66] for automatic term extraction.

Aligning extracted terms We used standard word alignment methods [2] to align the terms found in a pair of parallel sentences. We further imposed that the obtained term pairs should have similar-enough embeddings. Applying this method to pairs of parallel sub-corpora (C_1^j, C_2^j) provided a reference list $D_{1,2}^j$ of term translation pairs for each sub-corpus pair.

Sampling non-parallel corpora from parallel corpora A pair of parallel sub-corpora (C_1^j, C_2^j) can also be seen as a sub-corpus of aligned sentence pairs. We turned each such sub-corpus into a pair of non-parallel corpora ($C'_1{}^j$, $C'_2{}^j$) by discarding one of the two sentences in each sentence pair: the L_1 sentence was discarded with probability p_1 and the L_2 sentence with probability $1\text{-}p_1$.

Creating the input term lists and the gold standard for a pair of comparable sub-corpora Given a pair of sub-corpora ($C'_1{}^j$, $C'_2{}^j$), we obtained the input term list D_1^j for $C'_1{}^j$ as the list of terms found in the sentences in $C'_1{}^j$, and the input term list D_2^j for $C'_2{}^j$ as the list of terms found in the sentences in $C'_2{}^j$. We computed the gold standard dictionary $D'_{1,2}{}^j$ for ($C'_1{}^j$, $C'_2{}^j$) as the subset of term pairs (t_1, t_2) $\in D_{1,2}^j$ such that t_1 occurs in a sentence of $C'_1{}^j$ and t_2 occurs in a sentence of $C'_2{}^j$. At this point, the input term lists contain as many terms as can be found in the sentences of the input sub-corpora through a state-of-the-art ATE method. The gold standard list of term pairs contains the translations of these terms that were attested in the translations of these sentences and that happened to also occur in some other sentence kept in that sub-corpus.

5.9.2 Shared Task Results

Two teams (see Table 5.16) submitted runs and proposed several approaches for bilingual term alignment. The CUNI team [67] submitted three approaches: a cross-lingual embedding-based approach using the MUSE tool [37], a statistical phrase-based machine translation approach using the monoses pipeline [38] and contextualized embeddings from

Table 5.16 Shared task participants

System	Affiliation
CUNI	Institute of Formal and Applied Linguistics, Faculty of Mathematics and Physics, Charles University, Prague, Czech Republic [67]
IJS	Jožef Stefan Institute, Jožef Stefan International Postgraduate School, Ljubljana, Slovenia [68]

Table 5.17 Scores of the six submitted runs on the test set, in alphabetical order of run name: uninterpolated Average Precision, number of term pairs ranked by the system, number of gold standard term pairs, true positives, false positives, false negatives, precision, recall, F1-score. Bold shows the best result, underline the second best

run	AP	nSys	nGold	TP	FP	FN	P	R	F1
$CUNI_{\text{combined}}$	**0.2816**	6550	1970	945	5605	1025	0.1443	0.4797	0.2218
$CUNI_{\text{monoses}}$	0.1893	15477	1970	757	14720	1213	0.0489	0.3843	0.0868
$CUNI_{\text{muse}}$	0.1673	1570	1970	625	**945**	1345	**0.3981**	0.3173	**0.3531**
IJS_1	0.0054	14974	1970	366	14608	1604	0.0244	0.1858	0.0432
IJS_2	0.2674	54860	1970	**1576**	53284	**394**	0.0287	**0.8000**	0.0555
IJS_3	0.2685	54851	1970	1368	53483	602	0.0249	0.6944	0.0482

models such as multilingual BERT [23]. The IJS team [68] submitted three settings of an SVM-based machine learning approach that combines multiple features including sentence embeddings and dictionary-based features. Table 5.17 reports their results on the test set, after filtering out-of-vocabulary terms if necessary.

Overall, two of the IJS system settings (let us refer to them collectively as $IJS_{2,3}$) output considerably more suggestions, thus producing considerably more true positives, and a higher recall, but this comes at the expense of precision, which drives down both the AP and F1 scores. Conversely, two of the CUNI methods (MUSE and COMBINED) produce a much shorter list of term pairs which leads to a higher precision and F1-score.

Precision, recall, and F1-score are set measures that are computed on the whole set of returned term pairs, without taking their ranks into account. Conversely, average precision takes into account the ranks of the returned term pairs. Results are thus better analyzed by plotting the evolution of precision, recall and F1-score (Fig. 5.3) and average precision (Fig. 5.4) as term pairs are examined in the order in which they are ranked by a system. This shows that the $IJS_{2,3}$ methods, although returning a much longer list of term pairs than the $CUNI\backslash_{combined}$ method, rank true positives less close to the top of the list. Though they have a much higher recall, their average precision is slightly lower than that of $CUNI\backslash_{combined}$, which ranks more true positives closer to the top of the list.

According to the method in [69], the term pairs found by no systems are the most difficult examples and those found by all systems are the easiest examples. The most difficult exam-

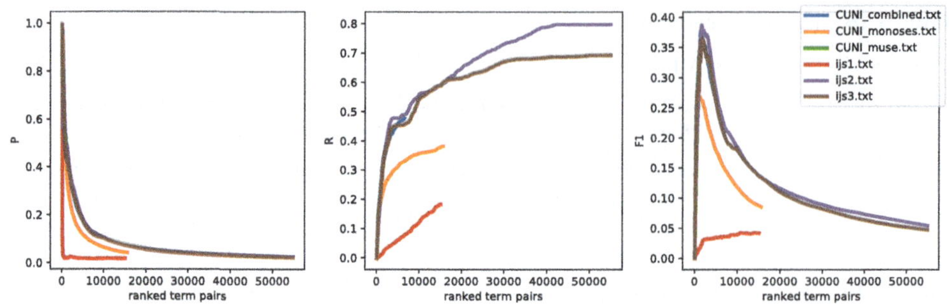

Fig. 5.3 Precision, recall, and F1-score against the number of examined term pairs

Fig. 5.4 Uninterpolated average precision against the number of examined term pairs

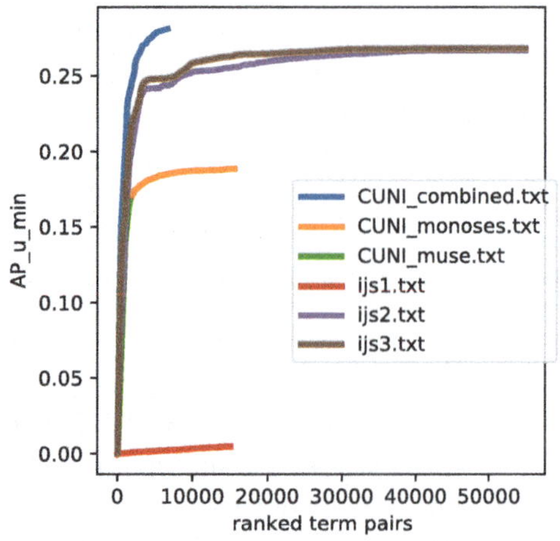

ples include for instance a polysemous word with three gold translations, none of which was found by any system, and a singular term whose gold translation is a singular term missed by all systems (a system proposed its plural instead). The easiest examples are all single-word terms, and source and target terms are identical or morphologically similar. These easiest examples hence display limitations similar to those in the MUSE dictionaries used in the BUCC 2020 shared task (see Sect. 5.8 above), but represent only a small part of the dataset.

The observed results show that the task of bilingual term alignment in comparable corpora, as defined and implemented through the BUCC 2022 dataset, is not yet solved. A limitation is that this shared task provided datasets for only one language pair: it needs to be extended to more languages that are typologically more distant from one another, an addition planned for the following years. Another limitation is the artificial nature of the corpus caused by its construction process, which is likely to differ from natural comparable corpora. This was a design choice meant to help control the presence of parallel terms and make an automatic evaluation possible.

References

1. Upadhyay S, Faruqui M, Dyer C, Roth D (2016) Cross-lingual models of word embeddings: an empirical comparison. In: Proceedings of the 54th annual meeting of the association for computational linguistics. Association for Computational Linguistics, Berlin, pp 1661–1670. https://doi.org/10.18653/v1/P16-1157. https://aclanthology.org/P16-1157
2. Brown PF, Cocke J, Della Pietra SA, Della Pietra VJ, Jelinek F, Lafferty J, Mercer RL, Roossin PS (1990) A statistical approach to machine translation. Comput Linguist 16(2):79–85
3. Philipp K (2005) Europarl: a parallel corpus for statistical machine translation. In: Proceedings MT Summit X
4. Gale WA, Church KW (1993) A program for aligning sentences in bilingual corpora. Comput Linguist 19(1):75–102. https://aclanthology.org/J93-1004
5. Och FJ, Ney H (2003) A systematic comparison of various statistical alignment models. Comput Linguist 29(1):19–51
6. Deerwester S, Dumais ST, Furnas GW, Landauer TK, Harshman R (1990) Indexing by latent semantic analysis. J Am Soc Inf Sci 41(6):391–407
7. Schutze H (1992) Dimensions of meaning. In: Proceedings of supercomputing '92. Minneapolis. pp 787–796
8. Rapp R (1995) Identifying word translations in non-parallel texts. In: Proceedings of the ACL. Cambridge, pp 320–322
9. Rapp R (1999) Automatic identification of word translations from unrelated English and German corpora. In: Proceedings of the 37th annual meeting of the association for computational linguistics. Association for Computational Linguistics, College Park, pp 519–526. https://doi.org/10.3115/1034678.1034756. https://aclanthology.org/P99-1067
10. Laroche A, Langlais P (2010) Revisiting context-based projection methods for term-translation spotting in comparable corpora. In: Proceedings of the 23rd international conference on computational linguistics (Coling 2010). Coling 2010 Organizing Committee, Beijing, pp 617–625. https://aclanthology.org/C10-1070
11. Sharoff S, Babych B, Hartley A (2006) Using comparable corpora to solve problems difficult for human translators. In: Proceedings of the international confenrence on computational linguistics and association of computational linguistics, COLING-ACL 2006. Sydney, pp 739–746
12. Irvine A, Callison-Burch C (2017) A comprehensive analysis of bilingual lexicon induction. Comput Linguist 43(2):273–310. https://doi.org/10.1162/COLI_a_00284. https://aclanthology.org/J17-2001
13. Faruqui M, Dyer C (2014) Improving vector space word representations using multilingual correlation. In: Proceedings of the 14th conference of the european chapter of the association for computational linguistics. Association for Computational Linguistics, Gothenburg, pp 462–471. https://doi.org/10.3115/v1/E14-1049. https://aclanthology.org/E14-1049
14. Gittins R (2012) Canonical analysis: a review with applications in ecology. Springer Science & Business Media
15. Baroni M, Dinu G, Kruszewski G (2014) Don't count, predict! A systematic comparison of context-counting versus context-predicting semantic vectors. In: Proceedings of the 52nd annual meeting of the association for computational linguistics. Baltimore
16. Goldberg Y (2017) Neural network methods for natural language processing, Synthesis lectures on human language technologies. Morgan & Claypool Publishers
17. Mikolov T, Yih W-T, Zweig G (2013) Linguistic regularities in continuous space word representations. In: Proceedings of the 2013 conference of the North American chapter of the association for computational linguistics: human language technologies. Association for Computational Linguistics, Atlanta, pp 746–751. https://aclanthology.org/N13-1090

18. Nissim M, van Noord R, van der Goot R (2020) Fair is better than sensational: man is to doctor as woman is to doctor. Comput Linguist 46(2):487–497. https://doi.org/10.1162/coli_a_00379. https://aclanthology.org/2020.cl-2.7
19. Raunak V, Kumar V, Gupta V, Metze F (2020) On dimensional linguistic properties of the word embedding space. In: Proceedings of the 5th workshop on representation learning for NLP. Association for Computational Linguistics, pp 156–165. https://doi.org/10.18653/v1/2020.repl4nlp-1.19
20. Vulić I, Ruder S, Søgaard A (2020) Are all good word vector spaces isomorphic? In: Proceedings of the 2020 conference on empirical methods in natural language processing (EMNLP). Association for Computational Linguistics, pp 3178–3192. https://doi.org/10.18653/v1/2020.emnlp-main.257
21. Dinu G, Lazaridou A, Baroni M (2014) Improving zero-shot learning by mitigating the hubness problem. arXiv preprint arXiv:1412.6568
22. Joulin A, Bojanowski P, Mikolov T, Jégou H, Grave E (2018) Loss in translation: Learning bilingual word mapping with a retrieval criterion. In: Proceedings of the 2018 conference on empirical methods in natural language processing. Association for Computational Linguistics, Brussels, pp 2979–2984. https://doi.org/10.18653/v1/D18-1330
23. Devlin J, Chang M-W, Lee K, Toutanova K (2019) BERT: Pre-training of deep bidirectional transformers for language understanding. In: Proceedings of the 2019 conference of the North American chapter of the association for computational linguistics: human language technologies, Volume 1 (Long and Short Papers). Association for Computational Linguistics, Minneapolis, pp 4171–4186. https://aclanthology.org/N19-1423
24. Anna R, Olga K, Anna R (2020) A primer in BERTology: what we know about how BERT works. Trans Assoc Comput Linguist 8:842–866
25. Miller GA, Leacock C, Tengi R, Bunker RT (1993) A semantic concordance. In: Proceedings of human language technology. Plainsboro. https://aclanthology.org/H93-1061
26. Peters ME, Neumann M, Iyyer M, Gardner M, Clark C, Lee K, Zettlemoyer L (2018) Deep contextualized word representations. In: Proceedings of the 2018 conference of the North American chapter of the association for computational linguistics: human language technologies. Association for Computational Linguistics, New Orleans, pp 2227–2237. https://doi.org/10.18653/v1/N18-1202. https://aclanthology.org/N18-1202
27. Coenen A, Reif E, Yuan A, Kim B, Pearce A, Viegas FB, Wattenberg M (2019) Visualizing and measuring the geometry of BERT. In: Advances in neural information processing systems, pp 8594–8603
28. Ethayarajh K () How contextual are contextualized word representations? comparing the geometry of BERT, ELMo, and GPT-2 embeddings. In: Proceedings of the 2019 conference on empirical methods in natural language processing and the 9th international joint conference on natural language processing (EMNLP-IJCNLP). Association for Computational Linguistics, Hong Kong, pp 55–65. https://doi.org/10.18653/v1/D19-1006. https://aclanthology.org/D19-1006
29. Biś D, Podkorytov M, Liu X (2021) Too much in common: shifting of embeddings in transformer language models and its implications. In: Proceedings of the 2021 conference of the North American chapter of the association for computational linguistics: human language technologies. Association for Computational Linguistics, pp 5117–5130. https://doi.org/10.18653/v1/2021.naacl-main.403
30. Cai X, Huang J, Bian Y, Church K (2021) Isotropy in the contextual embedding space: clusters and manifolds. In: International conference on learning representations
31. Hermann KM, Blunsom P (2014) Multilingual models for compositional distributed semantics. In: Proceedings of the 52nd annual meeting of the association for computational linguistics. Association for Computational Linguistics, Baltimore, pp 58–68. https://doi.org/10.3115/v1/P14-1006. https://aclanthology.org/P14-1006

32. Ormazabal A, Artetxe M, Labaka G, Soroa A, Agirre E (2019) Analyzing the limitations of cross-lingual word embedding mappings. In: Proceedings of the 57th annual meeting of the association for computational linguistics. Association for Computational Linguistics, Florence, pp 4990–4995. https://doi.org/10.18653/v1/P19-1492. https://aclanthology.org/P19-1492
33. Vulić I, Moens M-F (2013) A study on bootstrapping bilingual vector spaces from non-parallel data (and nothing else). In: Proceedings of the 2013 conference on empirical methods in natural language processing. Association for Computational Linguistics, Seattle, pp 1613–1624. https://www.aclweb.org/anthology/D13-1168
34. Mikolov T, Le QV, Sutskever I (2013) Exploiting similarities among languages for machine translation. arXiv preprint arXiv:1309.4168
35. Vulić I, Korhonen A (2016) On the role of seed lexicons in learning bilingual word embeddings. In: Proceedings of the 54th annual meeting of the association for computational linguistics. Association for Computational Linguistics, Berlin, pp 247–257. https://doi.org/10.18653/v1/P16-1024. https://aclanthology.org/P16-1024
36. Artetxe M, Labaka G, Agirre E (2016) Learning principled bilingual mappings of word embeddings while preserving monolingual invariance. In: Proceedings of the 2016 conference on empirical methods in natural language processing. Association for Computational Linguistics, Austin, pp 2289–2294. https://doi.org/10.18653/v1/D16-1250. https://aclanthology.org/D16-1250
37. Lample G, Conneau A, Ranzato M, Denoyer L, Jégou H (2018) Word translation without parallel data. In: Proceedings of the international conference on learning representations
38. Artetxe M, Labaka G, Agirre E (2019) An effective approach to unsupervised machine translation. arXiv preprint arXiv:1902.01313
39. Søgaard A, Vulić I, Ruder S, Faruqui M (2019) Cross-lingual word embeddings. Synthesis lectures on human language technologies. Morgan & Claypool Publishers
40. Artetxe M, Labaka G, Agirre E (2017) Learning bilingual word embeddings with (almost) no bilingual data. In: Proceedings of the ACL. Vancouver, pp 451–462
41. Alvarez-Melis D, Jaakkola T (2018) Gromov-Wasserstein alignment of word embedding spaces. In: Proceedings of the 2018 conference on empirical methods in natural language processing. Association for Computational Linguistics, Brussels, pp 1881–1890. https://doi.org/10.18653/v1/D18-1214
42. Søgaard A, Ruder S, Vulić I () On the limitations of unsupervised bilingual dictionary induction. In: Proceedings of the 56th annual meeting of the association for computational linguistics (volume 1: long papers). Association for Computational Linguistics, Melbourne, pp 778–788
43. Patra B, Moniz JR, Garg S, Gormley MR, Neubig G (2019) Bilingual lexicon induction with semi-supervision in non-isometric embedding spaces. In: Proceedings of the 57th annual meeting of the association for computational linguistics. Association for Computational Linguistics, Florence, pp 184–193. https://doi.org/10.18653/v1/P19-1018
44. Serge S (2020) Finding next of kin: cross-lingual embedding spaces for related languages. J Nat Lang Eng 26:163–182
45. Robert Ö, Jörg T (2016) Efficient word alignment with Markov Chain Monte Carlo. Prague Bull Math Linguist 106(1):125–146
46. Czarnowska P, Ruder S, Cotterell R, Copestake A (2020) Morphologically aware word-level translation. arXiv preprint arXiv:2011.07593
47. Huang J, Cai X, Church K (2020) Improving bilingual lexicon induction for low frequency words. In: Proceedings of the 2020 conference on empirical methods in natural language processing (EMNLP). Association for Computational Linguistics, pp 1310–1314. https://doi.org/10.18653/v1/2020.emnlp-main.100
48. Wu S, Conneau A, Li H, Zettlemoyer L, Stoyanov V (2020) Emerging cross-lingual structure in pretrained language models. In: Proceedings of the 58th annual meeting of the association

for computational linguistics. Association for Computational Linguistics, pp 6022–6034. https://www.aclweb.org/anthology/2020.acl-main.536
49. Conneau A, Lample G (2019) Cross-lingual language model pretraining. In: Conference on neural information processing systems. Vancouver, pp 7059–7069
50. Xu H, Koehn P (2021) Cross-lingual bert contextual embedding space mapping with isotropic and isometric conditions. arXiv preprint arXiv:2107.09186
51. Sennrich R, Haddow B, Birch A (2016) Neural machine translation of rare words with subword units. In: Proceedings of the 54th annual meeting of the association for computational linguistics. Berlin, pp 1715–1725. https://doi.org/10.18653/v1/P16-1162
52. Vernikos G, Popescu-Belis A (2021) Subword mapping and anchoring across languages. In: Findings of the association for computational linguistics: EMNLP 2021, Punta Cana, Dominican Republic. Association for Computational Linguistics, pp 2633–2647. https://doi.org/10.18653/v1/2021.findings-emnlp.224
53. Vulić I, Ponti EM, Litschko R, Glavaš G, Korhonen A (2020) Probing pretrained language models for lexical semantics. In: Proceedings of the 2020 conference on empirical methods in natural language processing (EMNLP). Association for Computational Linguistics, pp 7222–7240. https://doi.org/10.18653/v1/2020.emnlp-main.586
54. Glavaš G, Litschko R, Ruder S, Vulić I (2019) How to (properly) evaluate cross-lingual word embeddings: on strong baselines, comparative analyses, and some misconceptions. In: Proceedings of the 57th annual meeting of the association for computational linguistics. Association for Computational Linguistics, Florence, pp 710–721. https://doi.org/10.18653/v1/P19-1070
55. Hoshen Y, Wolf L (2018) Non-adversarial unsupervised word translation. In: Proceedings of the 2018 conference on empirical methods in natural language processing. Association for Computational Linguistics, Brussels, pp 469–478. https://doi.org/10.18653/v1/D18-1043
56. Kementchedjhieva Y, Hartmann M, Søgaard A (2019) Lost in evaluation: misleading benchmarks for bilingual dictionary induction. In: Proceedings of the 2019 conference on empirical methods in natural language processing and the 9th international joint conference on natural language processing (EMNLP-IJCNLP). Association for Computational Linguistics, Hong Kong, pp 3336–3341. https://doi.org/10.18653/v1/D19-1328
57. Sharoff S, Hartley A (2012) Lexicography, terminology and ontologies. In: Mehler A, Romary L, Gibbon D (eds) Handbook of technical communication (HAL 8). Mouton de Gruyter, Berlin, pp 317–346
58. Rapp R, Zweigenbaum P, Sharoff S (eds) Proceedings of the 13th workshop on building and using comparable corpora. European Language Resources Association, Marseille. ISBN 979-10-95546-42-9. https://aclanthology.org/2020.bucc-1.0
59. Marco B, Silvia B, Adriano F, Eros Z (2009) The WaCky wide web: a collection of very large linguistically processed web-crawled corpora. Lang Resourc Eval 43(3):209–226
60. Bojanowski P, Grave E, Joulin A, Mikolov T (2017) Enriching word vectors with subword information. Trans Assoc Comput Linguist 5:135–146. https://doi.org/10.1162/tacl_a_00051. https://www.aclweb.org/anthology/Q17-1010
61. Severini S, Hangya V, Fraser A, Schütze H (2020) LMU bilingual dictionary induction system with word surface similarity scores for BUCC 2020. In: Proceedings of the 13th workshop on building and using comparable corpora. European Language Resources Association, Marseille, pp 49–55. ISBN 979-10-95546-42-9. https://aclanthology.org/2020.bucc-1.8
62. Laville M, Hazem A, Morin E (2020) TALN/LS2N participation at the BUCC shared task: bilingual dictionary induction from comparable corpora. In: Proceedings of the 13th workshop on building and using comparable corpora. European Language Resources Association, Marseille, pp 56–60. ISBN 979-10-95546-42-9. https://aclanthology.org/2020.bucc-1.9

63. Adjali O, Morin E, Sharoff S, Rapp R, Zweigenbaum P (2022) Overview of the 2022 BUCC shared task: bilingual alignment in comparable specialized corpora. In: Proceedings of the 15th workshop on building and using comparable corpora. European Language Resources Association, Marseille, pp 67–76
64. Blancafort H, Daille B, Gornostay T, Heid U, Méchoulam C, Sharoff S (2010) TTC: terminology extraction, translation tools and comparable corpora. In: Proceedings of the EURALEX2010, Leeuwarden, 5–6 July 2010
65. Rocheteau J, Daille B (2011) TTC TermSuite - a UIMA application for multilingual terminology extraction from comparable corpora. In: Proceedings of the IJCNLP 2011 system demonstrations. Asian Federation of Natural Language Processing, Chiang Mai, pp 9–12. https://aclanthology.org/I11-2003
66. Cram D, Daille B (2016) Terminology extraction with term variant detection. In: Proceedings of ACL-2016 system demonstrations. Association for Computational Linguistics, Berlin, pp 13–18. https://doi.org/10.18653/v1/P16-4003
67. Požár B, Tauchmanová K, Neumannová K, Kvapiláková I, Bojar O (2022) CUNI submission to the BUCC 2022 shared task on bilingual term alignment. In: Proceedings of the 15th workshop on building and using comparable corpora. European Language Resources Association, Marseille, pp 43–49
68. Repar A, Koloski B, Ulčar M, Pollak S (2022) Fusion of linguistic, neural and sentence-transformer features for improved term alignment. In: Proceedings of the 15th workshop on building and using comparable corpora. European Language Resources Association, Marseille, pp 61–66
69. Gianola L, El Boukkouri H, Grouin C, Lavergne T, Paroubek P, Zweigenbaum P (2021) Differential evaluation: a qualitative analysis of natural language processing system behavior based upon data resistance to processing. In: Proceedings of the 2nd workshop on evaluation and comparison of NLP systems, Punta Cana, Dominican Republic. Association for Computational Linguistics, pp 1–10. https://aclanthology.org/2021.eval4nlp-1.1

6 Comparable and Parallel Corpora for Machine Translation

With the advent of Neural Machine Translation (NMT), a breakthrough has been achieved with regard to translation quality when compared to previous approaches such as rule-based, example-based, and statistical machine translation (MT). NMT systems tend to be considered as black boxes and it is not easy to predict their behavior. Nevertheless, the main reasons for the advances are probably the following two:

- NMT systems use an end-to-end-approach where all aspects of language, such as orthography, morphology, syntax, semantics and transfer from source to target language are taken into account simultaneously, thereby making it possible to take sophisticated interdependencies between processing levels into account.
- After decades of failure to resolve the problem, artificial neural networks finally made it possible to take long distance dependencies between language elements (e.g. between the words in a sentence) into account in the computer processing of natural language. They thus overcome a fundamental limitation of traditional n-gram-based approaches which dominated the field for decades.

On the downside, NMT systems are based on supervised learning and require large amounts of parallel corpora for training. Such corpora are usually available for pairs of the major world languages, but typically not for language pairs involving lesser used languages. This problem is especially severe as there are not many cases of parallel corpora where the same text has been translated into a large number of languages (an exception is e.g. the Bible). Therefore, if we wished to deal with only 1000 of the about 7000 world languages, this would result in $n^2 - n$, i.e. 999,000 language pairs and directions. As a parallel corpus

S. Sharoff et al., *Building and Using Comparable Corpora for Multilingual Natural Language Processing*, Synthesis Lectures on Human Language Technologies,
https://doi.org/10.1007/978-3-031-31384-4_6

can be used for training in both directions of a language pair, the acquisition of up to half of this number, i.e. 499,500 large parallel corpora for NMT training would be required. If we wished to specialize on particular text genres and domains, the number would further increase. Clearly, such numbers of parallel corpora cannot be acquired with reasonable effort.

This data acquisition bottleneck would be resolved to some extend if we had MT systems based on comparable rather than parallel corpora. One advantage of comparable corpora is that they are available in much larger quantities than parallel corpora as they do not require translation. Another advantage is that we would need only one corpus per language rather than per language pair, i.e. for 1000 languages 1000 monolingual corpora would be sufficient. Comparability could be ensured by looking, for example, at news texts which tend to report on similar incidents worldwide, or at the versions of Wikipedia in various languages where a lot of overlap between keywords (and consequently between the article contents) can be expected across languages.

An MT system based on comparable rather than parallel corpora would also have a higher cognitive plausibility as the human translation capability is clearly not based on learning from large amounts of parallel texts. Instead, by understanding text and speech, the human mind is capable to bridge the gap between languages. As this is the proof that there must be a way to learn translating without requiring parallel corpora, the challenge is to do so with computerized systems too, i.e. to build unsupervised MT systems.

In the following subsections we look at some of the better known previous attempts to achieve this goal.

6.1 MT Approach Based on Dictionaries Extracted From Comparable Corpora

For this approach we assume that we have a bilingual dictionary of inflected forms available for the language pair of interest. Such dictionaries could be induced from comparable corpora as described in Chap. 5, thus avoiding the need for a parallel corpus. We also need a large monolingual corpus of the target language. Given these resources, a short sentence can be translated by first looking up the words in the dictionary and generating a word-by-word translation. In a second step word order in the target language can be permuted, and the permutation with the highest occurrence frequency in the target language corpus can be considered to be the best translation.

As an example, let us look at the German sentence *dann passierte es* which is to be translated into English. A word-by-word translation would be *then happened it*, but as word order in English is different from German in this case, the expected translation would be *then it happened*. Let's suppose that a dictionary lookup in our German to English dictionary gives us the following word translations:

dann → then
passierte → happened, passed
es → it

The first and the last word are unambiguous, but *passierte* is ambiguous. Its two English translations covered in our dictionary relate to its main meanings which are *to pass through something* and *to happen* (there are further meanings but these are less common).

Potential word-by-word translations of the German sentence would be:

then happened it
then passed it

However, the second variant uses an inappropriate translation of the verb and both disregard English grammar which requires that the subject comes first and then the predicate follows. We can address this problem by systematically permuting word order as follows:

then happened it	then passed it
then it happened	then it passed
happened then it	passed then it
happened it then	passed it then
it happened then	it passed then
it then happened	it then passed

Now the idea is to look up the corpus frequency of each permutation in a large monolingual corpus of the target language and to then consider the most frequent permutation to be the correct translation. For the purpose of this example, instead of using a specific target language corpus we simply conducted a Google string search which has the advantage that it can be easily verified by the reader (though exact numbers can of course vary over time as the web evolves). At the time of writing this chapter, the Google search produced the frequencies as shown in Table 6.1.

As can be seen, in this example the expected translation obtained the highest frequency by a large margin. Note that in a real system doing a Google search may not be the best idea as using the Google API (application programming interface) is too slow for this specific purpose and as there are further drawbacks, among them interferences with other languages, and that punctuation and casing (upper and lower case characters) are ignored by Google.

Execution speed is critical as with increasing sentence length we run into a combinatorial explosion with regard to the number of possible permutations of word order. Therefore, even if we use fast algorithms, we start to run into problems for sentences longer than e.g. six words. Therefore, to apply the method to longer sentences, these should be split into shorter phrases. Hereby punctuation marks, conjunctions and prepositions can help to identify phrase boundaries. The phrases are then translated separately and the translation of the full sentence is obtained by concatenating the translations of the phrases.

Table 6.1 Google frequencies of word sequences

then it happened	24,800,000
then it passed	3,150,000
passed it then	1,660,000
happened then it	1,280,000
it then passed	1,100,000
then passed it	836,000
passed then it	699,000
happened it then	508,000
it happened then	353,000
it passed then	160,000
it then happened	99,000
then happened it	18,100

It should be emphasized that the simple method described above does not lead to results that are competitive with those of current NMT systems based on parallel corpora, which is why it has received only limited attention. A severe problem is that starting from a simple word-by-word translation does not account for cases where there is no matching terminology between the source and the target language and thus free (or even creative) translations are needed. Examples of this can be sophisticated multiword units whose overall meaning can deviate from what would be expected from the meanings of its component words, or metaphorical language use. The problem could be reduced by including sophisticated multiword units in the dictionary, but automatically extracting bilingual dictionaries of such multiword units is not an easy task on the basis of comparable corpora.

As this approach is fairly straightforward, it has been known for a long time. The above description is inspired by [1]. Probably the most advanced implementation of related work has been conducted by [2] who in this paper summarized and updated several previous studies, such as [3].

6.2 NMT Approach Using Parallel Corpora

As at the current state of the art there is no possibility to train an NMT system directly on comparable corpora, we focus in this section on parallel corpora which may or may not have been extracted from comparable corpora. In any case, as we consider comparable corpora a generic concept which also comprises parallel corpora, this is relevant in the framework of this book. The relevance is even higher as we will show strategies on how to perform multilingual NMT in a way where training is required for only a few language pairs, but where translations can be conducted for many more language pairs. Instead of substituting parallel by comparable corpora, which is often the goal of comparable corpora research, by utilizing recent findings we completely eliminate the need for most parallel corpora, thus pushing the boundaries of comparable corpora research.

As described in Chap. 4, parallel sentences can be extracted from comparable corpora. If the resulting parallel corpora are large enough, the standard statistical or neural machine translation methods can be applied successfully [4, 5].

Remember that extracting parallel sentences from comparable corpora can be conducted in mainly three ways: A straightforward one is to use a basic MT system to translate a source language corpus into the target language, and to identify similar sentences using a string similarity measure such as edit distance [6]. Problems with this approach are that a (though possibly very basic) MT system is required right from the beginning, and that MT is time consuming. The other two ways are based on embeddings. One way of using them is to obtain bilingual or multilingual word embeddings by learning them jointly from a parallel corpus. The other is to learn the embeddings for each language separately using monolingual corpora, and then to map them to a shared space based on a (possibly small) bilingual dictionary. Once we have the bilingual word embeddings, they can be extended to bilingual sentence embeddings. The task of finding sentence translations is then reduced to finding similar sentence embeddings across languages. See Chap. 4 for details.

Although only for the first method described above we need an MT system to identify sentences which are translations of each other, for all three of them we require an MT system to make use of the extracted parallel corpora. Both statistical and neural MT systems are well suited for this as they are data-driven. However, as neural systems usually promise better translation quality, we focus here on NMT.

As there is already a lot of literature on NMT [5, 7, 8]), let us give only a brief overview on how standard NMT works and how it utilizes parallel corpora, including ones that have been extracted from comparable corpora. NMT is not only relevant for the current considerations, but also for multilingual NMT as described later in this chapter.

6.2.1 Neural Networks for NMT

Standard neural networks can only deal with vectors of fixed lengths on the input and output layers. Unfortunately, this means that they cannot directly be used for MT. Hereby the problem is not to convert text into numbers, which is a straightforward task. A more serious obstacle is that text can be of arbitrary length. Although texts can be split into individual sentences, even sentences can consist of several hundred words. To process such lengthy input data, a very large neural network would be required, posing too high computational demands on the hardware. Fortunately, this problem could be solved using so-called *recurrent neural networks*. They focus on one word (or a few consecutive words) within a sentence at a time, and shift this focus in several steps until all words in the sentence have been processed. As indicated by the yellow arrows pointing backwards in Fig. 6.1, recurrent networks have a feedback loop from a network's output side to its input side. During processing, the previ-

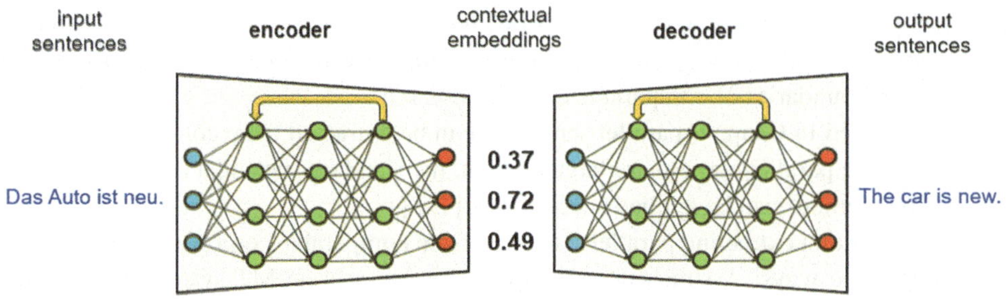

Fig. 6.1 Schematic diagram of the architecture of a neural MT system

ous foci are taken into account to some extend through these feedback loops, thus taking processing history (which here means context) into consideration. However, having such feedback loops across languages would go too far. Therefore, in current NMT systems the translation process is split into two parts corresponding to two neural networks, each with a separate feedback loop. The first network is called *encoder* and translates a source language sentence into a set of contextual vectors. The second one is called *decoder* and translates this set of contextual vectors into the target language. Figure 6.1 shows this well-established architecture.

For the purpose of focusing on particular parts of a sentence whilst translating it, as an alternative to recurrent neural networks it is also possible to use a so-called *attention mechanism* as provided in Google's *transformer* networks [9]. However, as going into the details of this is beyond the scope of this chapter, let us refer to the literature [7].

For training the network, a large parallel corpus with thousands or better millions of sentences paired with their translations is required. To start training, the first sentence of the source language is fed into the input layer of the encoder, and its translation (as taken from the parallel corpus) is used to compute the error when comparing it to what is generated at the output layer of the decoder. Using the back-propagation algorithm, the weights in the network are adjusted in small incremental steps. This is repeated for all sentence pairs in the parallel corpus, which concludes the first epoch. Typically, several epochs will be needed until the network converges, i.e. until further training does not significantly change the weights anymore.

After the training has completed, the network can be used in generation mode. In this mode, a new source language sentence (meant to be translated) is fed into the input layer of the encoder, and its translation is computed by the network. It is a general property of neural networks that they can generalize, i.e. if a source language sentence is fed in which has not been among the training sentences, it will still be translated, and usually the translation will be fairly good. Chances of a high quality translation are best if the input sentence is similar to sentences in the training corpus, i.e. shows similar statistical regularities. This is more likely if it belongs to the same genre and domain. Note, however, that it is difficult or even almost impossible to find out which statistical regularities the network detected during training as this information is hidden in the complex connection patterns between thousands or millions of neurons.

Translation quality of the system can be improved by adding some preprocessing with the aim of segmenting all words fed into the system into subword units of highly repetitive character sequences. For example, the German word *durchgegart*, meaning *cooked through*, could be split into *durch* (through), *ge* (participle marker), *gar* (cook) and *t* (marker for third person). However, in practice, rather than using sophisticated morphological analyzers to do the splitting, much simpler approaches have been shown to be highly effective. A well-established approach is *byte pair encoding* [10] which uses techniques similar to those known from file compression to identify highly repetitive subword units. It has been shown that byte pair encoding works best if its subword units are determined by taking together the source and target parts of the training corpus, i.e. the subwords are computed across languages.

Looking at subword units makes it possible that what the network learns is not limited to statistical regularities between the words of a sentence pair, but also considers regularities between the respective subword units. This way it can take not only syntactical and semantical but also morphological information into account. It turns out that by doing so the system can often make sense of previously unseen words that occur in new sentences to be translated, thereby to some extend resembling the capabilities of humans to deal with unknown words.

NMT systems as described above work very well in practice. They can be implemented using toolkits such as Marian NMT or FAIRSEQ. A step-by-step introduction on how to do this is, for example, provided in [8].

6.2.2 Reducing the Need for Parallel Corpora Using Multilingual NMT

The above descriptions of NMT assumed that we had one source and one target language. However, in the NMT framework it is easily possible to extend the number of languages on both the source and the target side. In the following subsections, let us look at this step by step.

1. *Several source languages, one target language*

First, we can extend the number of source languages but stay with one target language. As an example, Fig. 6.2 shows a system which is trained using parallel sentence pairs representing three source languages and one target language. Note that for training arbitrary sentences can be used, and that it is not essential that sentences with the same meaning but in different languages occur among the sentences on the source language side. As we deal with several languages on the source side, we need more information processing capacity especially in the encoder in comparison to the case with a single

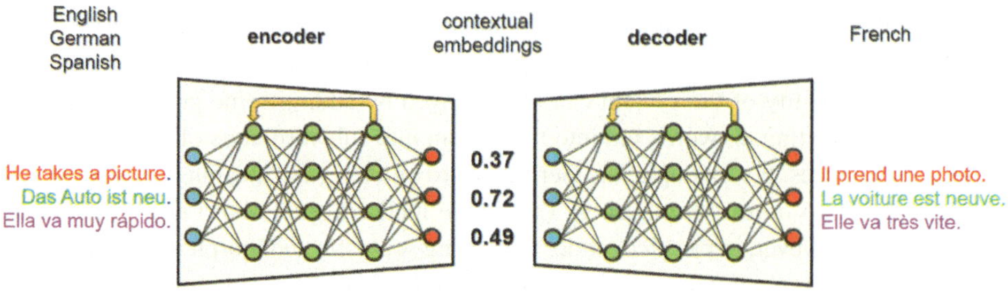

Fig. 6.2 Multilingual NMT system with three input languages and one output language

source language. Therefore, we should consider increasing the size of the neural network making up the encoder.

Normally it is not necessary to explicitly tell the encoder what source language it is dealing with, especially if the source languages in question are sufficiently distinct. However, if they are similar, problems with ambiguity in implicit language detection can occur. To give an example, let us assume that we have English and German as our source languages and that we wish to translate the phrase *"Made in Camembert"* into whatever target language. The problem here is that this phrase is perfectly valid in both English and German, but with rather different meanings. Taken as English, it says that something was manufactured in the French village of Camembert. However, if this phrase is taken to be German, the word *Made* has an entirely different meaning, standing for *maggot*. Therefore, the meaning of the sentence would likely be that a maggot has been found in a piece of Camembert cheese (or, less likely, that the maggot was found in the village of Camembert). Although typically it will not be necessary, we can avoid such confusions by adding a language marker to each of the source language sentences. In the example of Fig. 5.3, the input sentences could be modified to "[en] He takes a picture.", "[de] Das Auto ist neu" and "[es] Ella va muy rápido". As language markers, any out-of-vocabulary strings should be fine. Their placement could quite as well be at the end of the sentences. However, placing them somewhere in the middle of the sentences appears less appropriate as this could make it harder for the neural network to detect the statistical regularities (e.g. the word co-occurrence patterns) in the sentence.

Let us now consider how networks with several source languages can be used to improve low-resource translation. As an example, let us assume that we wish to translate from Catalan to German, but that we have only a small parallel corpus for training. Let us further assume that we have a large parallel corpus from Spanish to German, such as the Europarl corpus. We can now train the network with both corpora. Doing so is likely to significantly improve our translation results for Catalan to German in comparison to the case when we only train the system with the small Catalan to German corpus. The reason is that Catalan and Spanish are closely related languages, and therefore training with es-de also helps ca-de as e.g. many subword units are identical in both languages and show a similar behavior. On the other hand, if we translate with this mixed system from Spanish to German, the results may be slightly worse than with a pure es-de system for the reason that the parallel data for es-de is already of a good size, and that the additional ca-de data makes it somewhat noisier from the Spanish perspective. That is, in a multilingual system where a high- and a low-resource language are combined, the low-resource language tends to gain in translation quality, and the high-resource language tends to lose quality. The latter effect may be small if very large models are used.

2. *One source language, several target languages*

Having seen that a multilingual NMT system with several input languages and one output language is perfectly feasible, let us now turn to the opposite case, namely a multilingual NMT system with one input language und several output languages as shown in Fig. 6.3. In this case we have the problem that when translating we somehow have to tell the system which of the three target languages we want it to produce. This can be done using so-called *forced beginning of sentence* (forced BOS). This means that the decoder needs to have an option which allows to force it that the first words it comes up when starting to translate a sentence belong to the desired target language. If this can be ensured, the decoder will automatically produce the rest of the sentence in the same language. Forced BOS requires that this feature is supported by the particular decoder which is not always the case.
However, if the decoder does not support forced BOS, there is another possibility to achieve the same goal of determining the desired target language which works with any decoder. It is shown in Fig. 6.4 and uses language markers. However, other than described in the comments on Fig. 6.2 where the language markers indicated the source language, this time the language markers indicate the target language of a sentence pair used during training. In generation mode, when a new sentence is to be translated, simply the language marker of the desired target language needs to be added to this sentence. This should ensure that the decoder automatically comes up with the desired target language.

3. *Several source languages, several target languages*

Having looked at systems going from one to several and from several to one language, the next logical step is to go from several source languages to several target languages. This is shown in Fig. 6.5 for the simplest case of two source and two target languages. As in the previous case, we are using markers to determine the desired target language. In principle, this system could be trained with parallel corpora corresponding to the following language pairs: en-fr, en-de, es-fr, and es-de and good results could be expected as long as the neural networks in the encoder and the decoder are large enough to deal with the amount of information that comes with four rather than e.g. two language pairs.
However, as we now want to introduce the topic of *zero-shot translation*, let us assume that—as shown in Fig. 6.5, we only trained the system with the language pairs en-fr, es-fr and es-de. Obviously, after training the system will be able to translate from English to

French, from Spanish to French, and from Spanish to German. However, what happens now if, in generation mode, we feed the system with an English sentence that has attached to it a German target language marker? This is a situation which never occurred during training. The surprising finding, which was first discovered by a Google team [11], is that the system is able to translate from English to German although it was never trained with sentences from this language pair. This is called *zero-shot translation* because translation is possible on first attempt, with the amount of specific training being zero. As may be expected, in practice the translation quality in zero-shot translation is not as good as with explicit training of the respective language pair. Still, the question comes up, how is this possible at all?

To get an idea, let us look at the training sentence pairs shown in Fig. 6.5, namely "He takes a picture. / Il prend une photo." and "Él toma una foto. / Il prend une photo". In this case, we have an English and a Spanish input sentence which happen to have the same meaning and therefore lead to the same French translation. The easiest way to get the same translation for two different sentences is if the encoder comes up with identical contextual embeddings for both, i.e. if the embeddings generated by the encoder are language-agnostic. This is exactly what seems to happen in the system. Apparently, at the interface between encoder and decoder we get contextual embeddings which seem to represent the meaning of the input sentences. Such language-agnostic embeddings make it easy for the decoder to come up with the same translations for sentences with the same meaning and therefore have an advantage over other types of representations. This means that the training process is likely to give preference to such representations.

Note that in the above example a similar effect could also be achieved if we simply used French as an interlingual between English and German, and applied a traditional two-step translation procedure from English to French, and then from French to German. However, as will be shown in the next subsection on evaluation, zero-shot-translation also works when we train with es-en and fr-de, and then generate fr-en. In this case, the training pairs have no language in common and, surprisingly, zero-shot-translation nevertheless works to some extend, although not quite as well as in the example above.

Finally, let us come to another interesting application, namely a zero-shot grammar corrector. Let us assume that in Fig. 6.5 we added English to the target languages, and trained the system with en-fr, es-fr and en-de and es-de. What happens now if we ask the system to translate from English to English by entering, for example, the sentence "[en] He go fast."? Despite its grammatical shortcoming, the encoder is likely to "understand" the sentence, i.e. will come up with a language agnostic contextual representation of its meaning. Subsequently, the decoder will translate this representation into correct English for the reason that it was never trained to produce incorrect English. That is, our translation from English to English will be "He goes fast" which means that our system has corrected the English grammar without ever having been trained for this! In practice, further improvements on this are possible via so-called modularized NMT where encoders and decoders can be split into parallel sections representing particular language families (e.g. romanic, germanic, or slavic languages), see [12].

4. *Evaluation of zero-shot translation in multilingual NMT systems*

To give an idea of the performance of multilingual NMT systems as described above, we provide in Table 6.2 BLEU scores for some sample language pairs. The rightmost column of the table shows the BLEU scores of a conventional NMT system (transformer architecture) which was trained with the Europarl corpus of the language pair shown in column 2. These values are given as a baseline for comparison. More interesting are the BLEU scores in column 3. They were obtained when a multilingual system was trained with the language pairs listed in column 1, but when during generation the language pair listed in column 2 was enforced using target language markers in the source sentences. For example, when the system is trained on es-en and fr-de, generation for fr-en achieves a BLEU score of 34.11. In comparison, when the same system is trained on fr-en, generation for the same language pair achieves a BLEU score of 39.75 which is more than 5 BLEU points better. However, the score of 34.11 is still fairly good, and it is surprising that the system can achieve such a score for an unseen language pair (Fig. 6.3).

The rest of the table shows performances when the multilingual system is trained with three language pairs. In this case the scores get closer to the performance of the standard system, with discrepancies of only about 3 BLEU points on average. The results show that multilingual systems have a high potential in reducing the need for parallel corpora. The table also shows that the principles underlying multilingual systems do not require an interlingua. For example when training with es-en and fr-de, the two language pairs do not share a common language, i.e. there is no interlingua.

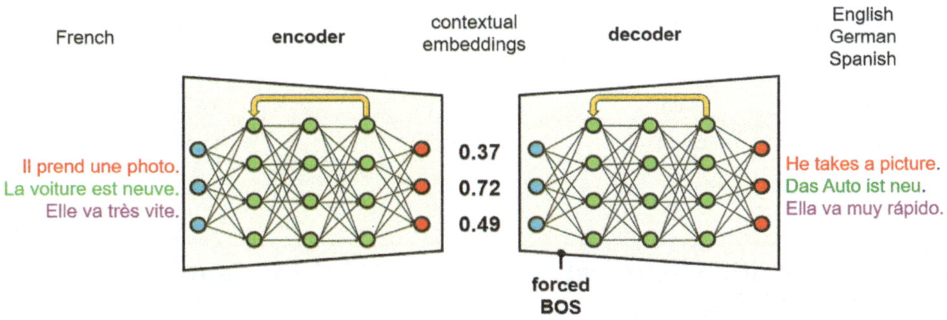

Fig. 6.3 Multilingual NMT system with one input language, three output languages and forced BOS (beginning of sentence)

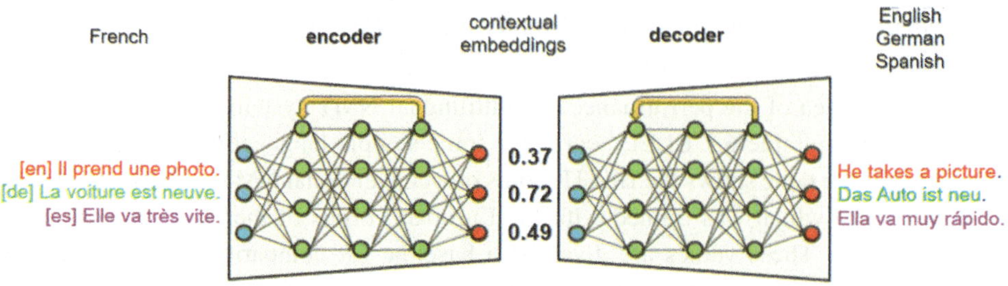

Fig. 6.4 Multilingual NMT system with one input language, three output languages and language markers indicating the target language in the source language sentences

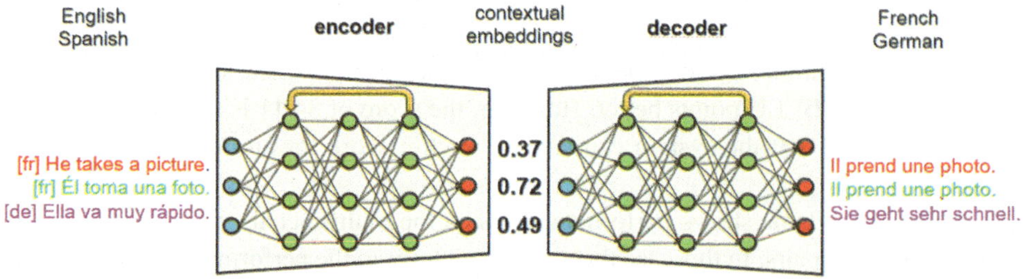

Fig. 6.5 Multilingual NMT system with two input languages, two output languages and language markers indicating the target language in the source language sentences

Table 6.2 comparison of BLEU scores of multilingual and standard NMT systems

Training	Translation	BLEU multilingual	BLEU standard
es-en, fr-de	fr-en	34.11	39.75
de-fr, es-en, es-fr	de-en	34.16	37.29
el-de, fr-en, fr-de	el-en	40.04	44.28
en-es, fr-de, fr-es	en-de	26.46	29.68
es-de, es-en, fr-de	fr-en	37.97	39.75
es-de, fr-en, fr-de	es-en	40.50	43.59

6.3 Information Retrieval Approach to MT Using Sentence Embeddings

As word embeddings are supposed to capture the semantics of words in the form of a vector, sentence embeddings are meant to capture the semantics of sentences. Like the meanings of sentences, ideally sentence embeddings should be language independent, that is, an original sentence and its translation into another language should have the same or at least similar sentence embeddings.

If we had a way to compute such sentence embeddings, MT could be considered as an information retrieval problem: First we would compute the sentence embeddings of all sentences in a very large target language corpus. To determine the translation of a source language sentence, we would also compute its sentence embedding. This we would compare to the sentence embeddings of all target language sentences using a standard vector similarity measure such as the cosine coefficient, and we would consider that target language sentence to be the translation of the original source language sentence whose embedding is most similar. The problem of combinatorial explosion could be dealt with in analogy to what was described in Sect. 6.1: We could split up the sentences into short phrases, retrieve their translations, and recombine the partial translations.

However, whereas there exist fairly established methods on how to compute word embeddings (see Chap. 5), it is not as clear how to do this for sentence embeddings and this is still a very active field of research. As discussed in Sect. 5.6, among the many suggestions how to do this, two main directions can be distinguished: One is to look at only a monolingual corpus, to compute word embeddings based on it, and to combine the monolingual word embeddings in various ways in order to form sentence embeddings. A problem with this approach is that the resulting sentence embeddings are by default not similar across languages and therefore, as for bilingual dictionary induction, mappings between languages are required which turn out to be problematic in practice.

Therefore, let us focus here on the second direction for computing sentence embeddings which, by starting from parallel corpora, uses a shared embedding space right from the beginning. At the current state of the art, such approaches seem better suited for MT research. If no parallel corpora are available to start with, they might be constructed by extracting parallel sentences from comparable corpora as shown in Chap. 4.

To give an example in analogy to the multilingual NMT systems discussed in Sect. 6.2.2, let us now consider a scenario were we have a multilingual language portfolio comprising ten languages plus English, and that we have available ten parallel corpora for each of the languages together with English. Using these parallel corpora, we could train ten standard NMT systems (as shown in Fig. 6.1) to translate from the ten languages into English. In principle we could also train another ten MT systems for the reverse directions, but this is not important for our following considerations. However, rather than using ten NMT systems, it is also possible to use only one system because NMT systems can easily deal with several languages, see Sect. 6.2.2. This is not only convenient, but—more importantly—also automatically creates a joint multilingual space for all languages.

Of course, if we wish our NMT system to work with ten language pairs rather than with one, we need to make sure that the neural network can deal with the larger amount of information by increasing its size (numbers of neurons and connections). Fortunately, however, we do not need to increase the size by a factor of ten as the information processing capability increases exponentially with the number of neurons (and connections) in the same way as the number of integers increases exponentially if we add digits to a number.

Therefore, it should be ok to increase the number of neurons by about the logarithm of the number of language pairs.

The neural network will be able to keep apart all language pairs if we add markers indicating the respective language to each source and target language sentence. E.g. the English sentence *"Then it happened"* could be marked *"[EN] Then it happened"*, and the German sentence *"Dann passierte es"* could be marked *"[DE] Dann passierte es"*. The exact strings used for marking and their positions within the sentences (e.g. at the beginning or at the end) are flexible if there is consistency across sentences. As long as the markers are unique (i.e. not occurring in the vocabulary) the neural network will be able to learn the details automatically.

However, if we mark all sentences by their language, this has two drawbacks: One is that the network will not be able to draw inferences between closely related languages such as Spanish, Catalan and Portuguese or between German and Dutch. Such inferences can help improving results when dealing with underresourced languages. The other, more important, drawback is that we want the encoder to come to identical representations at its output layer for sentences that are translations of each other. If we introduce language markers (especially on the source language side), this generalization across languages will likely not happen as the languages are kept separate by the markers. However, if we do not introduce the language markers, there is a chance that across languages sentences with similar meanings will arrive at similar encoder output representations (which we can take as sentence embeddings).

But why exactly can we expect similar representations at the output layer? The reason has to do with network economy. If two source sentences which are translations of each other lead to identical representations at the encoder output layer, it is guaranteed that they will have the same translation in the target language after passing the decoder, i.e. the purpose of the training process will definitely be fulfilled. However, if they end up having different representations at the output layer of the encoder, it is more difficult for the decoder to come up with the same (correct) translation into the target language English. Therefore a network that encodes similar sentence meanings across languages in similar output layer representations of the encoder is likely to have an advantage, i.e. should reach better optima during training.

Experimental work has shown that the quality of the generated language independent sentence embeddings can be further improved if not one but two target languages are used. As described in Sect. 6.2.2, in this case each of the target language sentences should be marked with one of the two target language identifiers, whereas adding markers indicating the source remains unnecessary. However, using two target languages of course means that parallel corpora from all source languages into two rather than one target language are required. Fortunately, we still don't need the full range of parallel corpora for all pairs of language combinations.

Once the network has been trained, for producing sentence embeddings, we do not need the decoder anymore as the embeddings are directly available at the output layer of the encoder. This saves computational effort as long as we are only interested in these embeddings and do not wish to use the system for end-to-end translation. This is an advantage in comparison to a system that simply considers the English target language translations as a form of sentence embeddings. Note that leaving out the decoding (after training is finished) also has the advantage that this error-prone processing step, which introduces noise to the sentence embeddings, is avoided.

Now that we have a system capable of generating language independent sentence embeddings, the question is what to do with it. As mentioned at the beginning of this section, one application could be to conduct machine translation from an information retrieval perspective. That is, for translating a source language sentence we compute its sentence embedding, and then we compare it to the sentence embeddings of all sentences in a very large target language corpus. The sentence with the most similar embedding is considered the best translation. Unfortunately, as the number of possible sentences in a language is astronomically high, it is rather unlikely that we will be able to find a reasonably good translation this way. The longer the source language sentence to be translated, the lower are the chances. This problem is so severe that systems using such a design are not used in practice. However, as described in Sect. 6.1, a solution could be to split longer sentences into phrases, to translate the phrases individually, and then to recombine the phrase translations, which is of course error-prone. This is a major drawback of such a system.

On the other hand, a nice feature of this system is that, although it required only ten parallel corpora all involving English, it is capable to translate between all pairs of our ten plus one languages, i.e. between $n^2 - n = 111$ language pairs and directions. (Note, however, that we would have the same advantage conventionally if the ten parallel corpora were translations of the same text, which is, however, not often the case in practice.)

Another natural application for our system would be to identify parallel sentences in comparable corpora. As in this case we are only interested in source language sentences which actually have a translation on the target language side of a comparable corpus, we avoid the problem of combinatorial explosion when trying to find sentence translations, and thus don't require splitting a sentence into sub-phrases. As also mentioned in Sect. 4.6, [23] very successfully used a system based on these principles called LASER (Language-Agnostic Sentence Representations) for extracting parallel sentences from comparable corpora. Although the extracted sentence pairs tend to be noisier than sentence pairs translated by humans, the authors used the resulting parallel corpora successfully to train conventional NMT systems, with fairly good results. On the webpage https://ai.facebook.com/blog/laser-multilingual-sentence-embeddings/ Holger Schwenk points out that the system has three important advantages:

- Low-resource languages and dialects benefit from the joint training especially with related languages from the same language family.
- The model supports code switching, i.e. the use of several languages in the same sentence.
- The performance of the system improves as more languages are added because it appears to learn characteristics within language families and possibly also universal

characteristics across languages. Therefore the representations could be seen as a universal language in a semantic vector space.

Note that there are many more interesting applications for a system capable of computing high-quality multilingual sentence embeddings, including grammar checking (see Sect. 6.2.2), paraphrasing and looking for sentences with similar meaning in the same or in different languages. If a classifier is trained on gold standard datasets such as the *Stanford Natural Language Inference Corpus*, such a system can also be used with good success on the task of *natural language interference* (NLI) which is sometimes also called *recognizing textual entailment*. NLI can be seen as an attempt to measure natural language understanding by machine. Given a sentence or sentence fragment called *premise*, the purpose of NLI is to determine whether a hypothesis relating to the premise is true (entailment), false (contradiction) or neutral (undetermined). The outcome neutral is to be chosen in case a decision cannot be made for lack of information. For example, given the premise "The fan is watching the soccer game" the hypothesis "The fan is a stone" should be judged to be false. The hypothesis "The fan has eyes" should be judged true, and the hypothesis "The fan is a woman" should be judged neutral. An advantage of a multilingual system for computing sentence embeddings as described above is that the premise and the hypothesis need not be in the same language.

As a final thought, let us come back to the previously mentioned idea of considering translations from any source language into an interlingua (in practice probably often a high-resource language such as English) as sentence embeddings. If the available MT systems work reasonably well, the aim of having similar embeddings for sentences with similar meanings would be fulfilled in this case, and the similarity between embeddings could be computed using measures for string comparison such as edit distance. However, when using standard NMT the training processes for each language pair would be separate and would not be using a shared multilingual space. Consequently, the training would not be designed to optimize multilingual embeddings in the way as described above, and thus the quality of the computed embeddings is likely to be worse. Also, the computational effort would be higher because the decoding phase could not be skipped, and because it requires more sophisticated computing to compare string similarities (e.g. via edit distance) rather than vector similarities (e.g. via the cosine coefficient). On the other hand, benefiting from two rather than one target language when generating sentence embeddings would be possible e.g. by concatenating the embeddings of the two languages.

6.4 Other Unsupervised MT Approaches

In this section we will try to give a concise overview on other MT approaches not requiring parallel corpora. For further details, we will provide pointers to the original literature. We also recommend to look at an overview article on recent advances in unsupervised NMT by

[13], an article describing the current limitations of unsupervised NMT [14] and a rather comprehensive survey of low-resource MT [15].

Statistical decipherment: Following [16], this approach [17, 18] considers machine translation to be a decipherment problem, assuming that a text in a foreign language (e.g. German) should be considered a cipher of the corresponding text in an original language (e.g. English). The decipherment is conducted as follows: First, individual words are deciphered by either applying an iterative version of the Expectation Maximization (EM) algorithm or by using Bayesian inference. To produce actual translations, a generative process using a standard n-gram language model is employed. In later work, it was shown that the results can be improved by incorporating syntactical information or word embeddings [19]. On the positive side, this approach does not require any parallel data. However, at the current state of development it is not competitive with standard NMT approaches in terms of translation quality.

Approach using cross-lingual embeddings, denoising and backtranslation: As described in Chap. 5, cross-lingual embeddings can be obtained from monolingual embeddings by aligning the vector spaces using mapping methods such as VecMap [20].

Although the cross-lingual embeddings obtained by mapping the vector spaces are not as good as the ones obtained using parallel corpora (see previous section), they were still good enough to encourage researchers to move on towards fully unsupervised NMT. To give a prominent example, let us summarize the approach implemented by [22]. It does not require any parallel data but only draws on mapped cross-lingual embedding spaces. It uses the standard NMT architecture as shown in Fig. 6.1 which, however is modified in three important ways:

- Similar to what we described in the previous section, the same system is used for both directions of a language pair.
- The same encoder is used for the source and target language. It is realized as a two-layer bidirectional recurrent neural network. This encoder is supposed to come up with language independent representations for both input languages. (It should be noted that [23] suggested a similar system which, however, not only uses a shared encoder but also a shared decoder.)
- The encoder uses fixed pre-trained cross-lingual embeddings which are kept fixed during training. Nevertheless, to avoid collisions between homographs (different words with the same spelling), the vocabularies for the two languages are kept separate. To give an example of a homograph, *die* in English means *decease*, but *die* in German is a definite article which translates to English *the*.

Due to the shared encoder, the system can be used to replicate its own input at the output in both directions. However, to avoid training results that blindly copy the input to the output, some noise is introduced, which is supposed to be removed by the system. As this denoising strategy turns out not to be sufficient, the system also uses backtranslation as suggested by [24]. This way pseudo-parallel sentence pairs are produced.

With regard to translation results, the system is claimed to achieve BLEU scores of 15.65 for French-English and 10.21 for German-English. These scores are considerably lower than the scores achieved based on parallel data which typically would be well beyond 30 for German-English and beyond 40 for French-English. However, this is nevertheless a remarkable achievement given the difficulty of the task.

Similar work has been conducted by the same authors using phrase-based statistical MT rather than NMT [25]. In this case an n-gram language model and a distortion model are employed. The underlying reasoning is that statistical MT can perform better than NMT in low resource settings and when dealing with noisy data such as the synthetic data produced in backtranslation.

Low-resource NMT: Given that fully unsupervised NMT is a very hard task, let us mention some ideas how to reduce rather than abandon the need for parallel data. One possibility is to go via a pivot language which makes sense if there is little parallel data when going directly from language A to language B, but if there is much more data when going from A to C and from C to B.

In another scenario, if for the language pair A1 to B there is only little parallel data, but if there exists a closely related language A2 with more data available from A2 to B, then it can help improve results if the NMT system is trained with both types of data in parallel.

As described in [23], it can also be beneficial to backtranslate a large target-language corpus, thus creating a synthetic parallel corpus. Starting from a basic system, this can help improving translation quality, and the process of backtranslation can be repeated to improve results iteratively. To reduce the danger of idiosyncratic behavior, this can be combined with training to replicate input sentences at the output.

References

1. Rapp R, Vide CM (2007) Statistical machine translation without parallel corpora. In: Datenstrukturen für linguistische Ressourcen und ihre Anwendungen/data structures for linguistic resources and applications. Proceedings of the Biennial GLDV conference 2007). Gunter Narr Verlag, Tübingen, pp 231–240
2. Irvine A, Callison-Burch C (2016) End-to-end statistical machine translation with zero or small parallel texts. Nat Lang Eng 22(4):517–548. https://doi.org/10.1017/S1351324916000127
3. Skadina I, Aker A, Mastropavlos N, Su F, Tufis D, Verlic M, Vasiljevs A, Babych B, Clough P, Gaizauskas R, Glaros N, Paramita ML, Pinnis M (2012) Collecting and using comparable corpora for statistical machine translation. In: Proceedings of the eighth international conference on language resources and evaluation (LREC'12). European Language Resources Association (ELRA), Istanbul, pp 438–445. http://www.lrec-conf.org/proceedings/lrec2012/pdf/925_Paper.pdf
4. Koehn P (2010) Statistical machine translation. Statistical machine translation. Cambridge University Press,

5. Koehn P (2020) Neural machine translation. Cambridge University Press. ISBN 9781108497329. https://books.google.gr/books?id=mdDqygEACAAJ
6. Levensthein VI (1966) Binary codes capable of correcting deletions, insertions, and reversals. Sov Phys Doklady 10(8):707–710. https://nymity.ch/sybilhunting/pdf/Levenshtein1966a.pdf
7. Kenny D (ed) Machine translation for everyone. Number 18 in translation and multilingual natural language processing. Language Science Press, Berlin. https://doi.org/10.5281/zenodo.6653406
8. Rapp R, How to build your own high-quality neural machine translation system using marian nmt. tcworld magazine, in print
9. Vaswani A, Shazeer N, Parmar N, Uszkoreit J, Jones L, Gomez AN, Kaiser Ł, Polosukhin I (2017) Attention is all you need. In: Advances in neural information processing systems
10. Sennrich R, Haddow B, Birch A, Neural machine translation of rare words with subword units. In: Proceedings of the 54th annual meeting of the association for computational linguistics. Berlin, pp 1715–1725. https://doi.org/10.18653/v1/P16-1162
11. Johnson M, Schuster M, Le QV, Krikun M, Wu Y, Chen Z, Thorat N, Viégas F, Wattenberg M, Corrado G, Hughes M (2017) Google's multilingual neural machine translation system: Enabling zero-shot translation. Trans Assoc Comput Linguist 5:339–351. https://doi.org/10.1162/tacl_a_00065. https://aclanthology.org/Q17-1024
12. Lyu S, Son B, Yang K, Bae J (2020) Revisiting modularized multilingual NMT to meet industrial demands. In: Proceedings of the 2020 conference on empirical methods in natural language processing (EMNLP). Association for Computational Linguistics, pp 5905–5918. https://doi.org/10.18653/v1/2020.emnlp-main.476
13. Biesialska M (2021) Major breakthroughs in... unsupervised neural machine translation (v). https://mt.cs.upc.edu/2021/03/08/major-breakthroughs-in-unsupervised-neural-machine-translation-v/
14. Kim Y, Graça M, Ney H (2020) When and why is unsupervised neural machine translation useless? CoRR. https://arxiv.org/abs/2004.10581
15. Haddow B, Bawden R, Barone AV, Helcl J, Birch A (2021) Survey of low-resource machine translation. https://www.researchgate.net/publication/354310341_Survey_of_Low-Resource_Machine_Translation
16. Weaver W (1949) Translation. https://repositorio.ul.pt/bitstream/10451/10945/2/ulfl155512_tm_2.pdf
17. Ravi S, Knight K (2011) Deciphering foreign language. In: Proceedings of the 49th annual meeting of the association for computational linguistics: human language technologies. Association for Computational Linguistics, Portland, pp 12–21. https://aclanthology.org/P11-1002
18. Dou Q, Knight K (2012) Large scale decipherment for out-of-domain machine translation. In: Proceedings of the 2012 joint conference on empirical methods in natural language processing and computational natural language learning. Association for Computational Linguistics, Jeju Island, pp 266–275. https://aclanthology.org/D12-1025
19. Dou Q, Vaswani A, Knight K, Dyer C (2015) Unifying Bayesian inference and vector space models for improved decipherment. In: Proceedings of the 53rd annual meeting of the association for computational linguistics and the 7th international joint conference on natural language processing (volume 1: long papers). Association for Computational Linguistics, Beijing, pp 836–845. https://doi.org/10.3115/v1/P15-1081
20. Artetxe M, Labaka G, Agirre E (2018) A robust self-learning method for fully unsupervised cross-lingual mappings of word embeddings. In: Proceedings of the 56th annual meeting of the association for computational linguistics (volume 1: long papers). Association for Computational Linguistics, Melbourne, pp 789–798. https://doi.org/10.18653/v1/P18-1073. https://aclanthology.org/P18-1073

21. Lample G, Conneau A, Ranzato MA, Denoyer L, Jégou H (2018) Word translation without parallel data. In: Proceedings of the international conference on learning representations
22. Artetxe M, Labaka G, Agirre E, Cho K (2018) Unsupervised neural machine translation. In: 6th International Conference on Learning Representations (ICLR 2018)
23. Artetxe M, Schwenk H (2019) Massively multilingual sentence embeddings for zero-shot cross-lingual transfer and beyond. Trans Assoc Comput Linguist 7:597–610. https://doi.org/10.1162/tacl_a_00288. https://aclanthology.org/Q19-1038
24. Sennrich R, Haddow B, Birch A (2016) Improving neural machine translation models with monolingual data. In: Proceedings of the 54th annual meeting of the association for computational linguistics (volume 1: long papers). Association for Computational Linguistics, Berlin, pp 86–96. https://doi.org/10.18653/v1/P16-1009. https://www.aclweb.org/anthology/P16-1009
25. Artetxe M, Labaka G, Agirre E (2018) Unsupervised statistical machine translation. arXiv preprint arXiv:1809.01272

Other Applications of Comparable Corpora 7

This section concerns applications of comparable corpora beyond pure machine translation. It has been argued [1, 2] that downstream applications such as cross-lingual document classification, information retrieval or natural language inference, apart from proving the practical utility of NLP methods, also provide a better assessment of the quality of multilingual vector spaces (and thus of comparable corpora used to build them) in comparison to intrinsic evaluation methods. For example, the results of intrinsic evaluation of embedding alignment methods are not a good predictor for the accuracy of cross-lingual document classification [2]. At the same time, given the same set of languages and the same neural architecture a cross-lingual space can be treated as more successful if it improves cross-lingual document classification.

7.1 Language Adaptation via Transfer of Language Resources

7.1.1 Zero/Few-Shot Transfer

One of the commonly used applications of comparable corpora concerns zero- or few-shot transfer of classifiers across languages. Usually this works in scenarios when a suitable training set is available for one or more donor languages L_{D_i}, while no (zero-shot) or very little (few-shot) data exists for a recipient language L_R. If we want to use a classifier for L_R, we can try building a cross-lingual vector space model which covers L_{D_i} as well as L_R, so that a classifier receives input in L_R, it is likely to make predictions for it in this VSM. A classifier in this case can aim at any task, for example, for text classification, question

S. Sharoff et al., *Building and Using Comparable Corpora for Multilingual Natural Language Processing*, Synthesis Lectures on Human Language Technologies,
https://doi.org/10.1007/978-3-031-31384-4_7

answering, natural language inference or for sequence-labelling tasks, such as POS tagging. Obviously, a small dataset in L_R is still needed for evaluating the transfer quality.

Both for traditional feature-based models and for neural representations, transfer learning assumes a semi-supervised procedure by inferring a vector representation space which takes into account both labeled "out-of-domain" data and unlabeled "in-domain" data [6, 7]. In cross-lingual setup this corresponds to treating L_D as an existing "out-of-domain" source and L_R as a new "in-domain" target. With the arrival of masked language models, very little explicit adaptation between L_D and L_R is needed apart from building the L_R vector space as similar to L_D. Comparability of corpora used for building the vector spaces is implicitly assumed by most scholars, but the impact has not been tested so far. For example, the multilingual BERT has been trained on comparable Wikipedias [8], while XLM-Roberta has been trained on samples from the Common Crawl [9].

We can expect a gap in performance between training with a native dataset in L_R and the transfer from L_{D_i} even if everything else is comparable between the training datasets in L_{D_i} and L_R, because the transfer might fail in fully transferring language-specific features. This gap can be measured, for example, by the transfer gap of suitable performance measures [10]:

$$TG(L_{D_i}, L_R) = Performance(L_{D_i}) - Performance(L_R)$$

With modern pre-trained transformer models, such as XLM-R, the gap of zero-shot transfer when L_D is English is below 0.25 (0.14 on average) on XTREME [10, 19], a standardised set of training resources for evaluation of language transfer measures on different tasks and language pairs, see individual examples below.

Zero-shot transfer via aligned cross-lingual VSMs was used with the traditional co-occurrence models, for example, by using cross-lingual n-gram-based spaces in text classification and terminology alignment [22]: if the two spaces are aligned, texts can be classified across languages. There has been proliferation of such studies starting from 2014 onwards, since neural embedding methods often provide higher quality spaces, which are nearly isomorphic across languages. This results in the possibility of a straightforward linear alignment of embeddings using a small seed dictionary [23]. Since neural classification models start with words represented as embeddings, zero-shot transfer can be achieved by direct application of models trained across languages. Contextual embedding models from pre-trained transformers, such as BERT or XLM-R, are even more efficient in building cross-lingual models, even in the absence of explicit supervision [9, 24].

The examples below focus on three tasks: Natural Language Inference (NLI), Named Entities Recognition (NER) and dependencies parsing. NLI is a 3-way classification task on the level of short segments, usually sentences, which should predict the relation for a pair of segments as entailment (the two statements are equivalent, $\leftrightarrow$), neutral (either statement can be true independently, N), and contradiction (only one statement can be true at the same time, $\not\leftrightarrow$), see examples in Table 7.1, taken from the development set of XNLI [25].

Table 7.1 Natural language inference relations

I don't have enough information	$\leftrightarrow$	I need to gather more information
I don't have enough information	N	I need more information to decide what kind of car to purchase
He turned in appeal to Lord Julian	$\not\leftrightarrow$	Lord Julian was nowhere to be seen

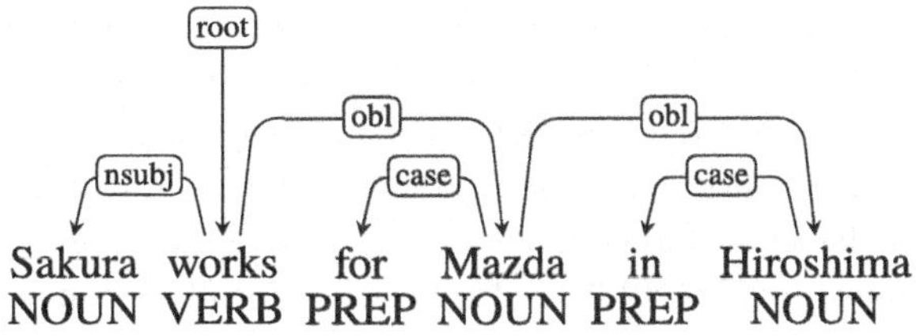

Fig. 7.1 Universal dependencies

NER is a sequence-labelling task which aims at giving each input token a label, such as PERson, LOCation, ORGanisation, as well as Other (for tokens which are not Named Entities). For example, the following sequence of annotations:

Sakura	*works*	*for*	*Mazda*	*in*	*Hiroshima*
PER	O	O	ORG	O	LOC

should resolve the ambiguities for *Sakura* or *Mazda*.

Finally, the dependencies parsing task is about detection of syntactic relations which link words in a sentence provided that the inventory of syntactic relations is shared across the L_D and L_R languages, as it is the case in the Universal Dependencies (UD) dataset [26], see an example of a dependency tree in Fig. 7.1.

The success in cross-lingual transfer of multilingual transformers is surprising, given the amount of linguistic knowledge needed for each application for each L_R language, such as the expectations of having good semantic knowledge for solving the NLI task, a good lexicon for the NER task and language-specific syntactic knowledge for the parsing task, for example, nouns typically come before adjectives in French, it is the other way around for English, while the models have been only trained on English.

Still the TG scores for cross-lingual transfer vary. This partly depends on the typological links between L_D and L_R (how closely the languages are related), as well as the scripts used for writing (for example, Latin, Cyrillic or Chinese). The typological and orthographic links become relevant in some tasks, but not in others. For example, the TG score using the XLM-R model trained in English is less than 0.07 for the NLI task for Romance or Slavonic language families, increasing to 0.18 for Sino-Tibetan or Niger-Congo languages [10]. At the same time, the TG score for the NER task ranges from negligible 0.01 for Romance languages, to 0.18 for Slavonic languages to 0.51 for Sino-Tibetan or Niger-Congo. Even though the use of a specific writing script is a superficial feature (many languages changed their scripts over the history of their cultural development), neural architectures, including the multilingual transformers, are affected by the script. Among other things they need to reduce their lexicons by using a Byte Pair Encoding (BPE) mechanism [28]. The BPE mechanism is very successful in operating with shorter lexicons without any explicit supervision. For example, the word form *webmasters* is not listed in the lexicon of BERT, but it can be split into BERT's BPE tokens as *web#master#s*, which is acceptable for many contexts of its processing. However, the BPE mechanism does not allow more complex transformations. For example, while the word *pharmacy* belongs to BERT's lexicon, the word *pharmacies* is split into BPE tokens as *ph#arm#ac#ie#s*, each of which is represented by a separate vector in the transformer output. It has been shown that languages not using the standard Latin alphabet tend to undergo more extensive BPE splitting through the lexicon of multilingual transformers [29]. See also Sect. 3.2 for the rationale for the BUCC shared task design in terms of the writing scripts and the typological variation of languages in comparable corpora.

Cross-lingual models are also reasonably successful in zero-shot transfer in non-topical classification tasks, such as genre classification or discourse prediction in spite of the fact that both genre traditions and discourse phenomena vary across languages and cultures. For example, the transfer gap for zero-shot prediction of Internet genres from English to Finnish using XLM-R is 0.11 (from 0.72 to 0.61 when measured in F1), while the monololingual F1 score when trained on Finnish corpora reaches only 0.73 [31]. Similarly, for prediction of discourse phenomena, such as the rhetorical structures, for example, detection of relations of contrast or exemplification, the average transfer gap is 0.13, while language dissimilarity has a modest effect, the accuracy of 0.62 for $L_D =$ English, 0.5 for German and 0.51 for Chinese [33].

The reason why the pre-trained transformers can align their representation spaces successfully without any supervision has been linked primarily to two things: (1) their ability to detect inherent structural similarities in these spaces while they learn their representations, and (2) the presence of anchor points, which are defined in this setting as *identical strings* in the respective lexicons.

The influence of the anchor points has been studied by removing them from the shared lexicon via adding a language prefix to each anchor point. So *London* becomes represented as *en:London* and *fr:London* in the lexicons for pre-training the transformers for English and French [9]. It is also possible to go in the opposite direction and to enrich the set of anchor points by adding entries from a bilingual dictionary. In this scenario, for each word pair $< w_D, w_R >$ listed in the dictionary, the comparable corpora used for pre-training are modified by replacing 30% of the respective words with their translations. Other experiments tried in [9] concerned the study of the contribution of different training corpora (Wikipedia for L_D vs Common Crawl for L_R) and different parameter sharing settings when the language models are trained separately in the initial layers in comparison to their joint training.

The results of these experiments, see Table 7.2, show that the most significant factor is the ability to build a shared representation which leads to detection of structural similarities

Table 7.2 Factors influencing the possibility of zero-shot transfer

Tasks:	NLI				NER				UD			
Languages:	fr	ru	zh	Δ	fr	ru	zh	Δ	fr	ru	zh	Δ
Default	73.6	68.7	68.3	0.0	79.8	60.9	63.6	0.0	73.2	56.6	28.8	0.0
No anchors	72.1	67.5	67.7	−1.1	74.0	57.9	65.0	−2.4	72.3	56.2	27.4	−0.9
Extra anchors	74.0	69.8	72.1	+1.8	76.1	59.7	66.8	−0.5	73.3	56.9	29.2	+0.3
Wiki-CC	74.2	65.8	66.5	−1.4	74.0	49.6	61.9	−6.2	71.3	54.8	25.2	−2.5
Sep Emb+L1-3	69.2	61.7	56.4	−7.8	73.8	46.8	53.5	−10.0	68.2	53.6	23.9	−4.3
Sep Emb+L1-6	51.6	35.8	34.4	−29.6	56.5	5.4	1.0	−47.1	50.9	6.4	1.5	−33.3

between the languages: the performance drops drastically when parameter sharing is disrupted and representation learning becomes monolingual in the lower layers. Removing or adding the anchor points is even less significant than the change of the underlying training corpus. Δ in Table 7.2 refers to the average difference of the respective performance measure from the default transfer setting.

This shows that the cross-lingual ability of pre-trained transformer models is primarily a by-product of their training procedure which involves sharing parameters across languages. However, this is achieved in the setting when enough data is available for each of the three L_R languages. It is expected that the transfer gap is much wider for severely under-resourced languages, when the amount of unlabelled raw-text training data is limited, see the results in [10, 34], as an under-resourced language does not have enough data to build a representation space which adequately reflects the properties of this language. The efforts to address the transfer performance gap in this case primarily focus on improving the lexical coverage and tokenisation, for example, via vocabulary augmentation and script transliteration. Vocabulary augmentation involves adding a small number of the most frequent BPE tokens from the target language (for example, 99) to the transformer lexicon and an additional round of pre-training on the target language. Transliteration involves conversion of the training corpora into the Latin script using standard tools, such as the `transliterate` package in Python. Both vocabulary augmentation and transliteration methods help to gain some performance, but usually this adds less than 1% [34], very much in line with the investigation of the influence of anchors in Table 7.2.

Most recently novel methods appeared which are geared towards learning suitable document-level embedding spaces directly. In particular, MARGE [35] is a pre-trained sequence-to-sequence model learned with an unsupervised multi-lingual multi-document paraphrasing objective. The objective noisily captures aspects of paraphrase, translation, multi-document summarization, and information retrieval, allowing for strong zero-shot performance on several tasks.

7.1.2 Language Adaptation for Related Languages

The language transfer methods discussed above do not make a distinction between the typological similarity of languages to close the transfer performance gap, as the training procedure remains the same for transfer from English to French or from English to Chinese. Some methods even emphasise the importance of abstracting from the specific features of the languages for training the embedding models, for example, by removing languages-specific means and variances in the embedding vectors and removing morphological contractions and reordering words, so that the adjectives always precede the nouns for all languages involved [36].

Some other models attach greater relevance to the typological links by assuming that there is specific information in closely related languages which can help in transferring the classifiers. One approach is to consider the amount of lexical similarity [37] assuming that embeddings for lesser-resourced languages can be improved by getting information from better-resourced languages in the process of aligning the cross-lingual embedding spaces. In particular it is possible to use the initial seed lexicon for building a character alignment model for detecting cognates. This alignment model can provide weights for Weighted Levenshtein Distance:

$$WLD(s_e, s_f) = \frac{\sum_{(e,f)\in al(s_e,s_f)}(1 - p(f|e))}{\max(len(s_e), len(s_f))}$$

where s_e and s_f are words in the two languages, al is a set of their alignments, $p(f|e)$ is the probability from the character alignment model. The distance is normalised by the length of the longest word. In the end, the word embedding aligned across related languages can be more successful in downstream tasks, such as morphological analysis, NER or genre classification, improving their performance by $\Delta = 0.23$, for example, in the genre classification task by transferring the classifiers from Russian into Ukrainian [37].

A different approach was tried for the task of MT Quality Estimation, which aims at predicting the amount of human post-editing efforts for MT output in the absence of reference translations. Classifiers for predicting MT quality for translations into Spanish have been transferred to predict quality for Italian and Portuguese [38]. This is achieved by Self-Taught Learning, a semi-supervised transfer learning method, which does not assume that the unlabelled dataset is drawn from the same distribution as the labelled one [39]. More specifically this involves the following steps:

1. Learn the dimensionality reduction for features of texts from the unlabelled sets z_i coming from both L_D and L_R using autoencoders;

2. Compute a new representation for the labelled training dataset x_i in L_D.
3. Use standard prediction methods for training a new model using a transformed training dataset $f(\hat{x}_i) = y_i$.
4. Apply the model to test sets across both L_D and L_R using $f(\hat{z}_i)$

This was shown to help in reducing language-specific variation and improves the isomorphism between the feature spaces across related languages, in the end reducing the transfer gap from 0.18 to 0.01 for the transfer from Spanish to Portuguese (the related measure is Pearson's correlation in a regression task).

7.2 Comparable Corpora in the Digital Humanities

This book mostly focused on applications of comparable corpora in Natural Language Processing tasks. In this section we will briefly sketch the many uses of comparable corpora in various areas of the Digital Humanities.

One area of their use concerns language teaching. The frequency lists derived from corpora have been providing a reliable source for designing language teaching curricula at least since the 1950s. However, an individual frequency list does not allow comparison with curricula in other languages. Also foreign language teaching is usually bilingual: the new language to be mastered against the native language of the learner. This situation requires comparison of frequency lists from several corpora and from several languages. A study reported in [40] used comparable corpora from a variety of languages, including four of the commonly taught languages (Arabic, Chinese, English and Russian) and five less commonly taught languages (Greek, Italian, Norwegian, Polish and Swedish). The aim was to compare the frequency lists of comparable corpora with respect to their utility for different levels of language learners. Cross-translation of the frequency lists helped in detecting what is common across languages and what is specific to individual cultures (such as food items). This also helped in assessing the degree of non-symmetric translation pairs, i.e. cases when a word like *rom* in Swedish has two meanings, 'rum' and 'roe', while each of them has only one translation into other languages. Overall, the translation symmetry depends on the typological similarity between the languages, but it is 24% on average with the highest value is for the English-Italian pair, down to 7.3% for the Arabic-Swedish pair.

In addition to the use of frequency lists derived from comparable corpora, foreign language learners themselves produce texts (for practicing or examination purposes), which provides a useful sour of information about their linguistic choices [41], in particular, when learners are given the task to produce controlled assignments in their respective languages they study, while they all have the same mother tongue. The errors shared across the languages can be matched to the levels of their linguistic proficiency.

Following the argument from the introduction of this book, linguistic research which compares linguistic phenomena across different languages cannot fully rely on parallel corpora, because of biases in topics, genres and translationese. Therefore, researchers rely on representative national corpora which can be treated as comparable resources as well [42], so that the frequencies in the use of, for example, discourse markers can be compared.

One of the recent efforts in corpus linguistics concerns development of The International Comparable Corpus [43, 44], which comes from well-controlled sources with respect to the domains and genres, so that this can lead to reliable linguistic generalisations in comparison to ad-hoc large collections, such as Wikipedias.

The same arguments about the biases of resources and translationese led to the active use of comparable corpora by professional translators [47], translation scholars [48], and translation trainers [49]. This provides us information about biased translation choices in parallel corpora. In particular, parallel corpora cannot provide the full picture for the use of ‘semantic prosody’ [51], which is the phenomenon of regular co-occurrences with a specific semantic set. For example, *to set in* can be defined *to become established*, but in text originally written in English it tends to co-occur with negative subjects, such as *bad weather* or *desperation*. However, this inference is possible only through comparable corpora, as its use in translations into English from other languages in parallel corpora is very infrequent, and when it is used as a translation, it does not exhibit the negative prosody.

As described in [52], comparable corpora are useful for translators, to get inspirations on how to solve translation problems, and in particular to confirm translation hypotheses for words or multiword units. In their daily work, translators often have the problem that in the source text special terms occur whose translations are not known. If available bilingual dictionaries or search engines for parallel texts such as linguee.com do not help, the standard method to find the translations of unknown technical terms is to look at texts of the target language which are similar in topic and content to the source text to be translated. Such comparable texts can be found, for example, by entering known translations of keywords characterizing the source text into a search engine, and by conducting an internet search. The translator then has to browse the comparable texts for passages where the missing translation occurs.

Comparable corpora also play an important role when it comes to investigating the properties of translated rather than original text. Often, the question arises whether the use of language in translations (the so called *translationese*) is different from the use of original texts. For example, the translators may leave their fingerprints, and it has often been claimed that the source language may shine through, e.g. the choice of vocabulary and grammatical constructions in the target language may be influenced by the source language. To study such phenomena, it can be useful to compare two corpora of the same language, one consisting of original texts and the other of translated texts [16].

However, comparable corpora are not limited to corpus linguistics and translation studies. In fact, there is a multitude of applications where they are useful when the researchers want to focus on implications arising from data. We can mention here a few examples typical for some fields:

- In comparative linguistics comparable corpora can be used to compare linguistic phenomena across languages and dialects.
- In diachronic linguistics language change can be studied over time. Here comparable corpora are particularly useful as the intuitions of current speakers do not well reflect language use of e.g. centuries ago.
- In discourse analysis, semantic fields, rhetorical features such as metaphor and deixis as well as discourse markers and discourse frames can be studied.
- In forensic linguistics texts can be attributed to authors by looking at (usually monolingual) comparable text collections of their publications.
- In pragmatics e.g. the effectiveness of advertisements in different languages can be studied.
- In sociolinguistics cultural differences can be studied e.g. across countries or communities.
- In typological research comparable corpora can be used to find out about language universals.

These uses of comparable corpora provide avenues for fruitful interaction between NLP rooted in computer sciences and the humanities.

References

1. Klementiev A, Titov I, Bhattarai B (2012) Inducing crosslingual distributed representations of words. In: Proceedings of the COLING, Mumbai
2. Glavaš G, Litschko R, Ruder S, Vulić I (2019) How to (properly) evaluate cross-lingual word embeddings: On strong baselines, comparative analyses, and some misconceptions. In: Proceedings of the 57th annual meeting of the association for computational linguistics. Association for Computational Linguistics, Florence, pp 710–721. https://doi.org/10.18653/v1/P19-1070. https://www.aclweb.org/anthology/P19-1070
3. Goldberg Y (2017) Neural network methods for natural language processing, Synthesis lectures on human language technologies. Morgan & Claypool Publishers, p 2017
4. Grégoire F, Langlais P (2018) Extracting parallel sentences with bidirectional recurrent neural networks to improve machine translation. In: Proceedings of the 27th international conference on computational linguistics. Association for Computational Linguistics, Santa Fe, pp 1442–1453. https://aclanthology.org/C18-1122
5. Grover J, Mitra P (2017) Bilingual word embeddings with bucketed CNN for parallel sentence extraction. In: Proceedings of ACL 2017, student research workshop. Association for Computational Linguistics, Vancouver, pp 11–16. https://aclanthology.org/P17-3003

6. Daumé H III, Kumar A, Saha A (2010) Frustratingly easy semi-supervised domain adaptation. In: Proceedings of the 2010 workshop on domain adaptation for natural language processing, Uppsala, pp 53–59. https://aclanthology.org/W10-2608
7. Bengio Y (2012) Deep learning of representations for unsupervised and transfer learning. In: Guyon I, Dror G, Lemaire V, Taylor G, Silver D (eds) Proceedings of ICML workshop on unsupervised and transfer learning. In: Proceedings of machine learning research, vol 27. PMLR, Bellevue, pp 17–36. https://proceedings.mlr.press/v27/bengio12a.html
8. Wu S, Dredze M (2020) Are all languages created equal in multilingual BERT? In: Proceedings of the 5th workshop on representation learning for NLP. Association for Computational Linguistics, pp 120–130. https://doi.org/10.18653/v1/2020.repl4nlp-1.16. https://aclanthology.org/2020.repl4nlp-1.16
9. Conneau A, Wu S, Li H, Zettlemoyer L, Stoyanov V (2020) Emerging cross-lingual structure in pretrained language models. In: Proceedings of the 58th annual meeting of the association for computational linguistics. Association for Computational Linguistics, pp 6022–6034. https://www.aclweb.org/anthology/2020.acl-main.536
10. Hu J, Ruder S, Siddhant A, Neubig G, Firat O, Johnson M (2020) XTREME: a massively multilingual multi-task benchmark for evaluating cross-lingual generalisation. In: Daum H III, Singh A (eds) Proceedings of the 37th international conference on machine learning. Proceedings of machine learning research, vol 119. PMLR, pp 4411–4421. https://proceedings.mlr.press/v119/hu20b.html
11. Huang E, Socher R, Manning C, Ng A (2012) Improving word representations via global context and multiple word prototypes. In: Proceedings of the 50th annual meeting of the association for computational linguistics (volume 1: long papers). Association for Computational Linguistics, Jeju Island, pp 873–882. https://aclanthology.org/P12-1092
12. Huang J, Cai X, Church K (2020) Improving bilingual lexicon induction for low frequency words. In: Proceedings of the 2020 conference on empirical methods in natural language processing (EMNLP). Association for Computational Linguistics, pp 1310–1314. https://doi.org/10.18653/v1/2020.emnlp-main.100. https://aclanthology.org/2020.emnlp-main.100
13. Irvine A, Callison-Burch C (2016) End-to-end statistical machine translation with zero or small parallel texts. Nat Lang Eng 22(4):517–548. https://doi.org/10.1017/S1351324916000127
14. Irvine A, Callison-Burch C (2017) A comprehensive analysis of bilingual lexicon induction. Comput Linguist 43(2):273–310. https://doi.org/10.1162/COLI_a_00284. https://aclanthology.org/J17-2001
15. Jakubina L, Langlais P (2016) BAD LUC@WMT 2016: a bilingual document alignment platform based on lucene. In: Proceedings of the first conference on machine translation: volume 2, shared task papers. Association for Computational Linguistics, Berlin, pp 703–709. https://doi.org/10.18653/v1/W16-2370. https://aclanthology.org/W16-2370
16. Jantunen J (2002) Comparable corpora in translation studies: strengths,and limitations. SKY J Linguist I5(43):105–117
17. Jegou H, Douze M, Schmid C (2011) Product quantization for nearest neighbor search. IEEE Trans Pattern Anal Mach Intell 33(1):117–128. ISSN 0162-8828. https://doi.org/10.1109/TPAMI.2010.57
18. Johnson J, Douze M, Jégou H. Billion-scale similarity search with GPUs. IEEE Trans Big Data 7(3):535–547. https://doi.org/10.1109/TBDATA.2019.2921572
19. Ruder S, Constant N, Botha J, Siddhant A, Firat O, Fu J, Liu P, Hu J, Garrette D, Neubig G et al (2021) XTREME-R: towards more challenging and nuanced multilingual evaluation. arXiv preprint arXiv:2104.07412
20. Sahlgren M, Holst A, Kanerva P (2008) Permutations as a means to encode order in word space. In: Proceedings of the 30th annual meeting of the cognitive science society. Washington DC

21. Salton G, Wong A, Yang C-S (1975) A vector space model for automatic indexing. Commun ACM 18(11):613–620
22. Forsyth R, Sharoff S (2014) Document dissimilarity within and across languages: a benchmarking study. Literary Linguist Comput 29:6–22
23. Mikolov T, Le QV, Sutskever I (2013) Exploiting similarities among languages for machine translation. arXiv preprint arXiv:1309.4168
24. Wu S, Dredze M (2019) Beto, bentz, becas: the surprising cross-lingual effectiveness of BERT. arXiv preprint arXiv:1904.09077
25. Conneau A, Lample G, Rinott R, Williams A, Bowman SR, Schwenk H, Stoyanov V (2018) XNLI: Evaluating cross-lingual sentence representations. In: Proceedings of the 2018 conference on empirical methods in natural language processing. Association for Computational Linguistics, Brussels, pp 2475–2485. https://doi.org/10.18653/v1/D18-1269. https://www.aclweb.org/anthology/D18-1269
26. Nivre J, de Marneffe M-C, Ginter F, Goldberg Y, Hajič J, Manning CD, McDonald R, Petrov S, Pyysalo S, Silveira N, Tsarfaty R, Zeman D (2016) Universal dependencies v1: a multilingual treebank collection. In: Proceedings of the tenth international conference on language resources and evaluation (LREC'16). European Language Resources Association (ELRA), Portorož
27. Och FJ, Ney H (2003) A systematic comparison of various statistical alignment models. Comput Linguist 29(1):19–51
28. Sennrich R, Haddow B, Birch A (2016) Neural machine translation of rare words with subword units. In: Proceedings of the 54th annual meeting of the association for computational linguistics. Berlin, pp 1715–1725. https://doi.org/10.18653/v1/P16-1162
29. Pfeiffer J, Vulić I, Gurevych I, Ruder S (2020) UNKs everywhere: adapting multilingual language models to new scripts. arXiv preprint arXiv:2012.15562
30. Pires T, Schlinger E, Garrette D (2019) How multilingual is multilingual BERT? arXiv preprint arXiv:1906.01502
31. Rönnqvist S, Skantsi V, Oinonen M, Laippala V (2021) Multilingual and zero-shot is closing in on monolingual web register classification. In: Proceedings of the 23rd nordic conference on computational linguistics (NoDaLiDa). Linköping University Electronic Press, Reykjavik, pp 157–165. https://aclanthology.org/2021.nodalida-main.16
32. Rose T, Stevenson M, Whitehead M (2002) The reuters corpus volume 1 -from yesterday's news to tomorrow's language resources. In: Proceedings of the third international conference on language resources and evaluation (LREC). European Language Resources Association (ELRA), Las Palmas, Canary Islands. http://www.lrec-conf.org/proceedings/lrec2002/pdf/80.pdf
33. Kurfalı M, Östling R (2021) Probing multilingual language models for discourse. arXiv preprint arXiv:2106.04832
34. Chau EC, Smith NA (2021) Specializing multilingual language models: an empirical study. arXiv preprint arXiv:2106.09063
35. Lewis M, Ghazvininejad M, Ghosh G, Aghajanyan A, Wang S, Zettlemoyer L (2020) Pre-training via paraphrasing. arXiv preprint arXiv:2006.15020
36. Zhao W, Eger S, Bjerva J, Augenstein I (2020) Inducing language-agnostic multilingual representations. arXiv preprint arXiv:2008.09112
37. Sharoff S (2020) Finding next of kin: cross-lingual embedding spaces for related languages. J Nat Lang Eng 26:163–182
38. Rios M, Sharoff S (2016) Language adaptation for extending post-editing estimates for closely related languages. Prague Bull Math Linguist 106:181–192
39. Raina R, Battle A, Lee H, Packer B, Ng AY (2007) Self-taught learning: transfer learning from unlabeled data. In: Proceedings of the ICML

40. Kilgarriff A, Charalabopoulou F, Gavrilidou M, Johannessen JB, Khalil S, Johansson Kokkinakis S, Lew R, Sharoff S, Vadlapudi R, Volodina E (2014) Corpus-based vocabulary lists for language learners for nine languages. Lang Resour Evaluat 48(1):121–163
41. Bański P, Gozdawa-Gołębiowski R (2010) Foreign language examination corpus for l2-learning studies. In: Proceeding of the workshop on building and using comparable corpora, Malta
42. Kupietz M, Witt A, Bański P, Tufiş D, Cristea D, Váradi T (2017) EuReCo – joining forces for a European reference corpus as a sustainable base for cross-linguistic research. In: Proceedings of the workshop on challenges in the management of large corpora and big data and natural language processing, Birmingham
43. Kirk J, Čermáková A, Oksefjell Ebeling S, Ebeling J, Kren M, Aijmer K, Benko V, Garabik R, Gorski R, Kupietz M, Jantunen J et al (2018) Introducing the international comparable corpus. In: Using corpora in contrastive and translation studies conference, Louvain. https://ids-pub.bsz-bw.de/frontdoor/deliver/index/docId/8248/file/Kirk_etal._Introducing_the_International_Comparable_Corpus_2018.pdf
44. Čermáková A, Jantunen J, Jauhiainen T, Kirk J, Křen M, Kupietz M, Dhonnchadha EU (2021) The international comparable corpus: challenges in building multilingual spoken and written comparable corpora. Res Corpus Linguist 9(1):89–103
45. Chen J, Nie J-Y (2000) Automatic construction of parallel English-Chinese corpus for cross-language information retrieval. In: Sixth applied natural language processing conference. Association for Computational Linguistics, Seattle, pp 21–28. https://doi.org/10.3115/974147.974151. https://aclanthology.org/A00-1004
46. Church K, Hanks P (1990) Word association norms, mutual information, and lexicography. Comput Linguist 16(1):22–29
47. Bernardini S, Ferraresi A (2013) Old needs, new solutions: comparable corpora for language professionals. In: Sharoff S, Rapp R, Zweigenbaum P, Fung P (eds) Building and using comparable corpora. Springer, pp 303–319
48. Xiao R (ed) (2020) Using corpora in contrastive and translation studies. Cambridge Scholars Publishing
49. Kunilovskaya M, Lapshinova-Koltunski E (2019) Translationese features as indicators of quality in English-Russian human translation. In: Proceedings of the human-informed translation and interpreting technology workshop (HiT-IT 2019). Incoma Ltd, Varna, pp 47–56. https://doi.org/10.26615/issn.2683-0078.2019_006. https://www.aclweb.org/anthology/W19-8706
50. Kunilovskaya M, Sharoff S (2019) Towards functionally similar corpus resources for translation. In: Proceedings of the international conference on recent advances in natural language processing (RANLP 2019). INCOMA Ltd, Varna, pp 583–592. https://doi.org/10.26615/978-954-452-056-4_069. https://aclanthology.org/R19-1069
51. Munday J (2011) Looming large: a cross-linguistic analysis of semantic prosodies in comparable reference corpora. In: Cruger A, Wallmach K, Munday J (eds) Corpus-based translation studies: research and applications. St Jerome Manchester, pp 169–186
52. Zanettin F (1998) Bilingual comparable corpora and the training of translators. Meta: Journal des traducteurs 43:12. https://doi.org/10.7202/004638ar

8 Conclusions and Future Research

In the beginning of the 2000s the use of comparable corpora was on the margins of NLP research. Existing MT systems were nearly always based on fully parallel corpora, while NLP applications were mostly built separately in each language without the advantages of cross-lingual transfer. The rise of neural methods, in particular the use of aligned word embeddings and more recently the development of multilingual pre-trained transformers, has moved comparable corpora to the center of NLP research, as multilingual language models are usually trained on comparable corpora. For example, mBERT [1] has been trained on Wikipedias, while XLM-Roberta [2] and mT5 [3] have been trained on Common Crawl [10]. The successful alignment of their embeddings leads to a large number of application areas, including low-resource MT or zero- and few-shot transfer of various classifiers across languages. This is the success story of comparable corpora. What is the future?

While making predictions in a fast-changing field is inherently difficult, we can try outlining several research directions in which we expect interesting progress. One concerns better understanding of the nature of neural representations in multi-dimensional vector spaces controlled by millions (recently billions) of parameters. Pre-trained language models are often presented as "foundation models" [12], because they encode the foundations for linguistic tasks, so that we can test that such models understand many phenomena without any explicit fine-tuning to recognize them [13]. However, we need to arrive at much better understanding of what these models know and what they do not know, and how to spot their infelicities. For example, when word embeddings are trained on Wikipedia, the median word hubness score (which measures to which extent a word is often in the k nearest neighbours of other words) is 5, while some words have a far larger score, such 156 for *cavalry*, which leads to their selection as translation candidates disproportionately more often [14]. What causes such infelicities? Can we find methods of repairing the vector spaces, either while we are building them or post-hoc?

S. Sharoff et al., *Building and Using Comparable Corpora for Multilingual Natural Language Processing*, Synthesis Lectures on Human Language Technologies,
https://doi.org/10.1007/978-3-031-31384-4_8

It is also important to know how similar the multilingual foundation models are across different languages. Empirically, the foundation models are successful in zero-shot transfer, but it is not clear how much foundation is shared across the languages: to what extent the multilingual BERT embeddings can be considered as an interlingual covering all languages? There is some emerging research on this topic, [15], but it is at its preliminary stage. A possible way of assessing such models is by prediction of how well a down-stream application can perform in zero-shot transfer [16]. Also we also need to acquire better understanding of how much foundation is transferred in zero-shot and few-shot transfer.

A related question concerns ways of building truly successful multilingual foundation models. Parallelism of representations in our current models has emerged because they share their parameters in the process of training a masked language model for several languages, just because of the need to have parsimonious representations covering all of the languages in the masked model. It noteable that such models do not improve much from any explicit supervision, such as anchor translations between the languages [2]. Linguists have some understanding of the similarities and differences between languages with various models often expressed through rules either shared or partitioned across languages [23]. How can this knowledge be used to improve the alignment of the embeddings and the transformer weights in multilingual neural models?

Another interesting research question concerns the amount of training data needed to achieve practical zero-shot transfer. The pre-trained models are successful when they consume a lot of data, far greater amounts in comparison to what humans need to learn a language or to prepare to a new task, for example, GPT-3 has been trained on a 500 billion word corpus, which should take about 90,000 years to read by humans. It has been shown that the cross-lingual spaces are weaker for lesser-resourced languages [24], primarily because their models have fewer parameters to share with better-resourced languages, so their representation space is incomplete and less aligned with better-resourced languages. There is nascent research into re-alignment of pre-trained models for the languages missing from training, for example [25], but so far such methods only concern the initial embedding layer of these models, rather than fine-tuning of their weights. Another approach involves the production of synthetic data, either through available crude machine translation or simply by word-to-word translation with a bilingual lexicon [30]. Can there be better methods for tuning the parameters for lesser-resourced languages? If the foundation models are truly foundational, can they provide good foundations for adaptation to lesser-resourced languages?

Finally, an open research question concerns researching the impact of different strategies for creating multilingual models from unrelated comparable corpora. The developers of XLM-Roberta and mT5 used different strategies to create "clean" subsets of the Common Crawl, while the two models are also based on different architectures. What is the impact of the difference in the architectures versus the difference in the degree of comparability of their constituent corpora? There is also no shortage of methods for extracting more comparable documents, see Chap. 3, as well as more comparable sentences, Chap. 4. Will their use impact the quality of multilingual models?

References

1. Devlin J, Chang M-W, Lee K, Toutanova K (2019) BERT: pre-training of deep bidirectional transformers for language understanding. In: Proceedings of the 2019 conference of the north american chapter of the association for computational linguistics: human language technologies, volume 1 (long and short papers). Association for Computational Linguistics, Minneapolis, pp 4171–4186. https://aclanthology.org/N19-1423
2. Conneau A, Wu S, Li H, Zettlemoyer L, Stoyanov V (2020) Emerging cross-lingual structure in pretrained language models. In: Proceedings of the 58th annual meeting of the association for computational linguistics. Association for Computational Linguistics, pp 6022–6034. https://www.aclweb.org/anthology/2020.acl-main.536
3. Raffel C, Shazeer N, Roberts A, Lee K, Narang S, Matena M, Zhou Y, Li W, Liu PJ (2020) Exploring the limits of transfer learning with a unified text-to-text transformer. J Mach Learn Res 21(140):1–67
4. Raina R, Battle A, Lee H, Packer B, Ng AY (2007) Self-taught learning: Transfer learning from unlabeled data. In: Proceedings of the ICML
5. Rapp R (1995) Identifying word translations in non-parallel texts. In: Proceedings of the ACL. Cambridge, pp 320–322
6. Rapp R (1999) Automatic identification of word translations from unrelated English and German corpora. In: Proceedings of the 37th annual meeting of the association for computational linguistics. Association for Computational Linguistics, College Park, pp 519–526. https://doi.org/10.3115/1034678.1034756. https://aclanthology.org/P99-1067
7. Rapp R (2004) A freely available automatically generated thesaurus of related words. In: Proceedings of the language resources and evaluation conference, (LREC'04). Lisbon, pp 395–398
8. Rapp R, How to build your own high-quality neural machine translation system using marian nmt. Tcworld magazine, in print
9. Rapp R, Vide CM (2007) Statistical machine translation without parallel corpora. In: Datenstrukturen für linguistische Ressourcen und ihre Anwendungen/data structures for linguistic resources and applications. Proceedings of the biennial GLDV conference 2007). Gunter Narr Verlag, Tubingen, pp 231–240
10. Dodge J, Sap M, Marasović A, Agnew W, Ilharco G, Groeneveld D, Mitchell M, Gardner M (2021) Documenting large webtext corpora: a case study on the colossal clean crawled corpus. In: Proceedings of the 2021 conference on empirical methods in natural language processing. Punta Cana, Dominican Republic. Association for Computational Linguistics, pp 1286–1305. https://doi.org/10.18653/v1/2021.emnlp-main.98. https://aclanthology.org/2021.emnlp-main.98
11. Dou Q, Knight K (2012) Large scale decipherment for out-of-domain machine translation. In: Proceedings of the 2012 joint conference on empirical methods in natural language processing and computational natural language learning. Association for Computational Linguistics, Jeju Island, pp 266–275. https://aclanthology.org/D12-1025
12. Bommasani R, Hudson DA, Adeli E, Altman R, Arora S, von Arx S, Bernstein MS, Bohg J, Bosselut A, Brunskill E et al (2021) On the opportunities and risks of foundation models. arXiv preprint arXiv:2108.07258
13. Anna R, Olga K, Anna R (2020) A primer in BERTology: what we know about how BERT works. Trans Assoc Comput Linguist 8:842–866
14. Serge S (2020) Finding next of kin: cross-lingual embedding spaces for related languages. J Nat Lang Eng 26:163–182
15. Del M, Fishel M (2021) Establishing interlingua in multilingual language models. arXiv preprint arXiv:2109.01207

16. Ahuja K, Anastasopoulos A, Patra B, Neubig G, Choudhury M, Dandapat S, Sitaram S, Chaudhary V (2022) The SUMEval 2022 shared task on performance prediction of multilingual pretrained language models. In: Proceedings of the first workshop on scaling up multilingual evaluation. Association for Computational Linguistics, pp 1–7. https://aclanthology.org/2022.sumeval-1.1
17. Alvarez-Melis D, Jaakkola T (2018) Gromov-Wasserstein alignment of word embedding spaces. In: Proceedings of the 2018 conference on empirical methods in natural language processing. Association for Computational Linguistics, Brussels, pp 1881–1890. https://doi.org/10.18653/v1/D18-1214. https://aclanthology.org/D18-1214
18. Antonova A, Misyurev A (2011) Building a web-based parallel corpus and filtering out machine-translated text. In: Proceedings of the 4th workshop on building and using comparable corpora: comparable corpora and the web. Association for Computational Linguistics, Portland, pp 136–144. https://aclanthology.org/W11-1218
19. Artetxe M, Schwenk H (2019) Margin-based parallel corpus mining with multilingual sentence embeddings. In: Proceedings of the 57th annual meeting of the association for computational linguistics. Association for Computational Linguistics, Florence, pp 3197–3203. https://doi.org/10.18653/v1/P19-1309. https://aclanthology.org/P19-1309
20. Artetxe M, Schwenk H (2019) Massively multilingual sentence embeddings for zero-shot cross-lingual transfer and beyond. Trans Assoc Comput Linguist 7:597–610. https://doi.org/10.1162/tacl_a_00288. https://aclanthology.org/Q19-1038
21. Artetxe M, Labaka G, Agirre E (2016) Learning principled bilingual mappings of word embeddings while preserving monolingual invariance. In: Proceedings of the 2016 conference on empirical methods in natural language processing. Association for Computational Linguistics, Austin, pp 2289–2294. https://doi.org/10.18653/v1/D16-1250. https://aclanthology.org/D16-1250
22. Artetxe M, Labaka G, Agirre E (2017) Learning bilingual word embeddings with (almost) no bilingual data. In: Proceedings of the ACL. Vancouver, pp 451–462
23. Bateman JA (1997) Enabling technology for multilingual natural language generation: the KPML development environment. Nat Lang Eng 3(1):15–55
24. Wu S, Dredze M (2020) Are all languages created equal in multilingual BERT? In: Proceedings of the 5th workshop on representation learning for NLP. Association for Computational Linguistics, pp 120–130. https://doi.org/10.18653/v1/2020.repl4nlp-1.16. https://aclanthology.org/2020.repl4nlp-1.16
25. Minixhofer B, Paischer F, Rekabsaz N (2022) WECHSEL: effective initialization of subword embeddings for cross-lingual transfer of monolingual language models. In: Proceedings of the 2022 conference of the north american chapter of the association for computational linguistics: human language technologies. Association for Computational Linguistics, Seattle, pp 3992–4006. https://doi.org/10.18653/v1/2022.naacl-main.293. https://aclanthology.org/2022.naacl-main.293
26. Moore RC (2002) Fast and accurate sentence alignment of bilingual corpora. In: Machine translation: from research to real users. In: 5th conference of the association for machine translation in the Americas. Springer, Heidelberg, pp 135–244
27. Munday J (2011) Looming large: a cross-linguistic analysis of semantic prosodies in comparable reference corpora. In: Cruger A, Wallmach K, Munday J (eds) Corpus-based translation studies: research and applications. St Jerome Manchester, pp 169–186
28. Munteanu DS, Marcu D (2005) Improving machine translation performance by exploiting non-parallel corpora. Comput Linguist 31(4):477–504
29. Munteanu DS, Marcu D (2006) Extracting parallel sub-sentential fragments from non-parallel corpora. In: Proceedings of the 21st international conference on computational linguistics and

44th annual meeting of the association for computational linguistics. Association for Computational Linguistics, Sydney, pp 81–88. https://doi.org/10.3115/1220175.1220186. https://www.aclweb.org/anthology/P06-1011

30. Wang X, Ruder S, Neubig G (2022) Expanding pretrained models to thousands more languages via lexicon-based adaptation. In: Proceedings of the 60th annual meeting of the association for computational linguistics (volume 1: long papers), Dublin, pp 863–877. https://doi.org/10.18653/v1/2022.acl-long.61. https://aclanthology.org/2022.acl-long.61